FEAR and Loathing in Introductory Biology

Thomas V. Vogel
Western Illinois University

Kendall Hunt
publishing company

Cover image courtesy of Chet Bain

Kendall Hunt
publishing company

www.kendallhunt.com
Send all inquiries to:
4050 Westmark Drive
Dubuque, IA 52004-1840

ISBN: 978-0-7575-6539-7

Printed in the United States of America
10 9 8 7 6 5

DEDICATION

I would like to dedicate this work to my wife and children, for all of the years of wonderful questions they have asked have made me keep looking for the answers.

Know that science never has and never will prove anything! The goal of science is to provide evidence that supports some claim. That does not mean it is right nor does that mean it is wrong. Only the best evidence we can provide at this time. Question everything and never stop looking for new evidence that supports a new view. That is the one thing that keeps science dynamic. And yes, *Lampyridae* is a family of insects in the beetle order Coleoptera. They are winged beetles, and commonly called fireflies or "lightning bugs" for their use of bioluminescence to attract mates. Tomorrow they may be classified as something different because some new piece of evidence is found to show they are more closely related to the "Flood Monkey's" of Cedar Rapids!

That being said:

> If knowledge is power . . . and power is strength . . . and strength is the opposite of weakness . . . and weakness breeds fear . . . and fear nurtures contempt . . . and contempt fuels hate . . . and hate begets war . . . therefore: knowledge brings Peace. So, by buying this book and taking my class we will bring about world peace! Awesome! And thank you for stopping by.

CONTENTS

PREFACE

This course is divided into six units. Each unit will have an exam that will summarize the material and allow for group study and provide opportunity for cooperative learning. All exams are in multiple-choice format and will be found online.

There will be no cumulative exam. The lecture portion of this class will make up 75% of the total grade and the remaining 25% will come from Lab. I will not be recording attendance, although I do expect you to attend Lecture whenever you can. I would estimate that roughly 70% of the material I expect you to know for each exam is in this text, with 25% coming from the stories and movies shown during class. The remaining 5% will come from outside sources and from the use of your very capable minds.

Introduction

Don't panic! Trust me when I tell you it will be fine. You are not the first class of students that will experience the angst of Intro Biology and I am certain you will not be the last. As you huddle in the comfort of the cinderblock room that will be "home" for the next four years (yes, Virginia, you can graduate in four years, although it did take me nine years for my Bachelor of Science!), remember to pace yourself. There is time enough for the full college experience. I would hate for you to miss anything that will make for a colorful story for your children and grandchildren. As this is an introductory course, we are forced to include a phenomenal amount of material. Because we cover so much, we will only be touching on each of the topics for a short time before we find ourselves moving on. Please feel free to ask as many questions as you wish. I will not have all of the answers. I do promise to do my best to find the most complete and timely answers to anything you ask—safe from prejudice, blame, fear, or failure. The more questions you ask, the more time I can devote to teaching and less time lecturing. That's right, my standing in front of you three days a week and talking is not the key to learning. Learning is a permanent change in behavior due to experience. How many times did your parents warn you "Don't touch that, it's hot," only to wait until no one was watching you so you could finally touch it. Don't be ashamed. Everybody had to do the same thing to learn what "hot" means. I do not see my role as your teacher—life is your teacher. I am just your storyteller. Hopefully, something you have heard in class will be there in your subconscious. When life sends you an experience, you will remember the story from the lecture and have that light bulb moment we live for. Learning is in the doing—lecturing to you is my way of giving you a preview, so when the experience comes, it is easier for you to make sense of it and learn from it. Or in the case of my less traditional students who may have witnessed many of these events already, this will help to clarify what happened or why. To my family and all overeager to learn: Thank you and hang on; we won't be back this way again!

UNIT ONE

When is science?

Who am I?

What is that smell?

Do these socks go with these pants?

WHERE DID I COME FROM?

HOW DID I GET HERE?

INTRODUCTION

EXAM I

Nature of Science
Metrics
Chemistry
Molecular Evolution

Characteristics of Life
Systems

- Acquire Material and Energy
- Homeostatic
- Response to Stimuli
- Reproduce
- Grow and Develop
- Adapt
- Evolve

Characteristics of Life

Levels of Organization

- Cell—the smallest level of organization that can be considered alive
- Tissue—groups of cells with similar structure and function
- Organ—biological machines made up of different tissues that carry out a particular function
- Organ System—made up of various organs that carry out major body functions

Levels of Organization

- Organism—made up of a finite number of organ systems
- Population—all of the members of a given species in a given place at a given time
- Community—all of the different species that occupy a given place at a given time
- Ecosystem—a community and the physical environment in which it lives
- Biosphere—narrow envelope surrounding the Earth, where all life is found

Scientific Method

- State the Problem
- Collect Known Information
- Formulate a Hypothesis
- Test the Hypothesis
- Observe and Record Result
- State a Conclusion
  1. Replicate
  2. Develop a Theory
  3. Publish

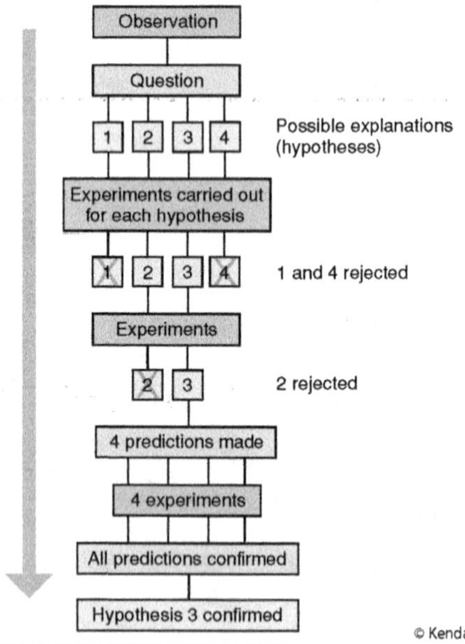

Nature of Science

Deductive vs. Inductive

- Deductive Reasoning—making individual decisions by applying accepted general principles
- Typical decision-making process

Nature of Science

Deductive vs. Inductive

- Inductive Reasoning—way of discovering general principles by careful examination of specific cases
- Became popular about 400 years ago (in Europe)
- Develop a hypothesis about how the world works
- Perform experiments to support the hypothesis

Deductive vs. Inductive

A	B

Scientific Investigation
Controlled Experiment

- To evaluate alternative hypothesis about one variable
- All other variables must be kept constant
- This eliminates misleading or confusing results

Scientific Investigation
Controlled Experiment

- **Control Group**—does not get the Experimental Factor
- **Experimental Group**—exactly the same as the Control Group—plus it gets the Experimental Factor

Scientific Investigation
Controlled Experiment

- Two experiments are done at the same time
- This gives the validity to the conclusion
- Helps to reduce Experimental Error

Sample Controlled Experiment

Experimental Group	Control Group

Metric System
Conversions

- **Kilo**—1000 times larger than the Base Unit
- **Base Unit**
- **Deci**—1/10th the size of the base unit
  - 1/10 = .1
- **Centi**—1/100th the size of the base unit
  - 1/100 = .01
- **Milli**—1/1000th the size of the base unit
  - 1/1000 = .001

Metric System
Conversions

- **Step 1**—Place what you know on the Top Left (Unit and Number)
- **Step 2**—Place the units you want to end up on the Top Right
- **Step 3**—Place the units you want to get rid of on the Bottom Right
- **Step 4**—Conversion Factor: The bigger unit always gets a one

Metric System
Conversions

- **Step 5**—Math
- **Cancel Units**
  - When the number on the bottom is one—multiply
  - When the number on the bottom is not one—divide by that number
- **Move the decimal right when you multiply**
  - Equal to the number of zeroes you are multiplying by
- **Move the decimal left when you divide**
  - Equal to the number of zeroes you are dividing by

Metric System

Conversions

- 35 liter = ? ml
- 35L—Top Left
- mL—Top Right
- L—Bottom Right
- L is the larger unit * it gets a one
- How much larger *1000 times, so the mL gets a 1 and three zeroes

Metric Conversion Problems

Example	Step 1	Step 2	Step 3	Step 4	Step 5
2,500 mL = ______l					
.0085 cm = ______m					
.00005 km = ______cm					

Metric Conversion Problems

Example	Step 1	Step 2	Step 3	Step 4	Step 5
500 mm = ______m					
.085 m = ______mm					
.000005 kg = ______g					

Metric Conversion Problems

Example	Step 1	Step 2	Step 3	Step 4	Step 5
450 dg = ______g					
.0003 l = ______cl					
1,000,000 mm = _______ km					

Chemistry

Atoms

- The smallest particle into which a substance can be divided and still retain its chemical properties
- Mass—refers to the amount of substance
  - Often used interchangeably with weight even though they have slightly different meanings
- Nucleus—small, very dense area found in the center of each atom

Subatomic Particles

Protons

- Found in the nucleus
- Atomic mass of +1
- Charge of +1
- Used as the identification unit for **Atomic Number**

Subatomic Particles

Electrons

- Found outside the nucleus in the electron cloud
- Atomic mass so small, we will call it Zero
- Charge of −1
- Equal to Energy

Energy Levels

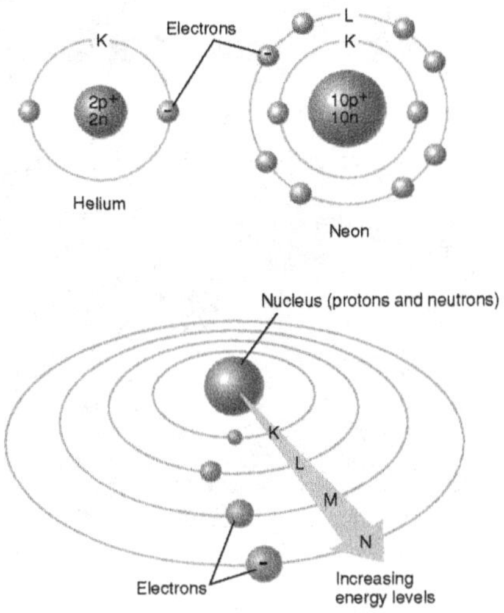

Subatomic Particles

Neutrons

- Found inside the nucleus
- Mass of +1
- Charge of zero (Neutral)

Atomic Models

Element	Atomic Structure	Lewis Dot Diagram	Bonding Rules	Bonding Model
Chlorine				
Carbon				
Hydrogen				

Atomic Models

Element	Atomic Structure	Lewis Dot Diagram	Bonding Rules	Bonding Model
Nitrogen				
Oxygen				
Sodium				

Chemistry
Terminology

- Atomic Number—equal to the number of protons in the nucleus of an atom
- Atomic Mass—equal to the number of protons and neutrons in the nucleus of an atom
- Element—any substance that cannot be broken down to any other substance through normal chemical means

Periodic Table

Name - Hydrogen
Atomic # - 1
Symbol - H
Atomic weight - 1.0079

- Alkali metals
- Alkaline earth metals
- Transition metals
- Rare earth metals
- Other metals
- Non-metals
- Halogens
- Noble (inert) gases

Period	Group IA	IIA	IIIB	IVB	VB	VIB	VIIB	VIIIB	VIIIB	VIIIB	IB	IIB	IIIA	IVA	VA	VIA	VIIA	VIIIA
1	Hydrogen 1 H 1.0079																	Helium 2 He 4.0026
2	Lithium 3 Li 6.941	Beryllium 4 Be 9.0122											Boron 5 B 10.811	Carbon 6 C 12.0112	Nitrogen 7 N 14.0067	Oxygen 8 O 15.9994	Fluorine 9 F 18.9984	Neon 10 Ne 20.179
3	Sodium 11 Na 22.989	Magnesium 12 Mg 24.305											Aluminum 13 Al 26.9815	Silicon 14 Si 28.086	Phosphorus 15 P 30.9738	Sulfer 16 S 32.064	Chlorine 17 Cl 35.453	Argon 18 Ar 39.948
4	Potassium 19 K 39.098	Calcium 20 Ca 40.08	Scandium 21 Sc 44.956	Titanium 22 Ti 47.90	Vanadium 23 V 50.942	Chromium 24 Cr 51.996	Manganese 25 Mn 54.938	Iron 26 Fe 55.847	Cobalt 27 Co 58.933	Nickel 28 Ni 58.71	Copper 29 Cu 63.546	Zinc 30 Zn 65.38	Gallium 31 Ga 69.723	Germanium 32 Ge 72.59	Arsenic 33 As 74.922	Selenium 34 Se 78.96	Bromine 35 Br 79.904	Krypton 36 Kr 83.80
5	Rubidium 37 Rb 85.468	Strontium 38 Sr 87.62	Yttrium 39 Y 88.905	Zirconium 40 Zr 91.22	Niobium 41 Nb 92.906	Molybdenum 42 Mo 95.94	Technetium 43 Tc (99)	Ruthenium 44 Ru 101.07	Rhodium 45 Rh 102.905	Palladium 46 Pd 106.4	Silver 47 Ag 107.868	Cadmium 48 Cd 112.40	Indium 49 In 114.82	Tin 50 Sn 118.69	Antimony 51 Sb 121.75	Tellurium 52 Te 127.60	Iodine 53 I 126.904	Xenon 54 Xe 131.30
6	Cesium 55 Cs 132.905	Barium 56 Ba 137.34	57	Hafnium 72 Hf 178.49	Tantalum 73 Ta 180.948	Tungsten 74 W 183.85	Rhenium 75 Re 186.2	Osmium 76 Os 190.2	Iridium 77 Ir 192.2	Platinum 78 Pt 195.09	Gold 79 Au 196.967	Mercury 80 Hg 200.59	Thalium 81 Tl 204.37	Lead 82 Pb 207.19	Bismuth 83 Bi 208.980	Polonium 84 Po (209)	Astatine 85 At (210)	Radon 86 Rn (222)
7	Francium 87 Fr (223)	Radium 88 Ra (226)	89	Rutherfordium 104 Rf (261)	Hahnium 105 Ha (262)	Seaborgium 106 Sg (263)	Neilsbohrium 107 Ns (264)	Hassium 108 Hs (265)	Meitnerium 109 Mt (266)									

Lanthanides	Lanthanum 57 La 138.91	Cerium 58 Ce 140.12	Praseodymium 59 Pr 140.907	Neodymium 60 Nd 144.24	Promethium 61 Pm 144.913	Samarium 62 Sm 150.35	Europium 63 Eu 151.96	Gadolinium 64 Gd 157.25	Terbium 65 Tb 158.925	Dysprasium 66 Dy 162.50	Holmium 67 Ho 164.930	Erbium 68 Er 167.26	Thulium 69 Tm 168.934	Yiterbium 70 Yb 173.04	Lutetium 71 Lu 174.97
Actinides	Actinium 89 Ac (227)	Thorium 90 Th 232.038	Protactinium 91 Pa (231)	Uranium 92 U 238.03	Neptunium 93 Np (237)	Plutonium 94 Pu 244.064	Americium 95 Am (243)	Curium 96 Cm (247)	Berkelium 97 Bk (247)	Californium 98 Cf 242.058	Einsteinium 99 Es (254)	Fermium 100 Fm 257.095	Mendelevium 101 Md 258.10	Nobelium 102 No 259.101	Lawrencium 103 Lr 260.105

Chemistry
Isotopes

- Atoms that have the same number of protons (Atomic Number) but have a different number of neutrons
- Therefore they have different Atomic Mass
- **Example:**
  - Carbon—12
  - Carbon—14

Chemistry
Molecules

- A group of atoms held together by energy
- Ionic Bonds—form when atoms exchange electrons with each other
  - *Ions* are atoms with an electrical charge, due to an unequal number of electrons and protons
  - Ions are attracted to ions of the opposite charge
- Covalent Bonds—form when atoms share electrons with each other

Ionic Chemical Bonding

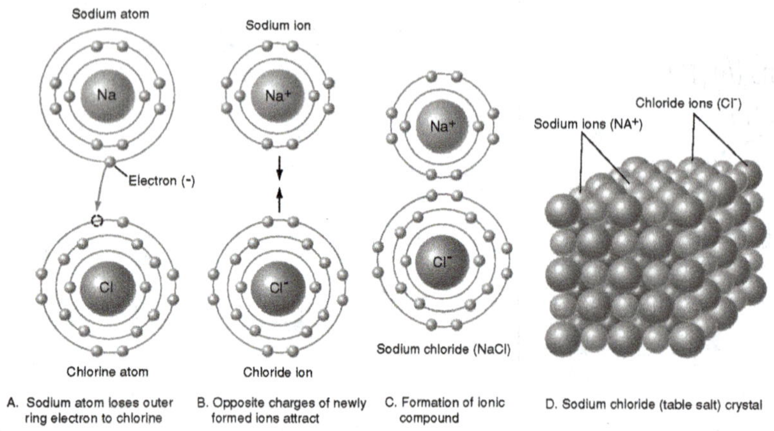

Covalent Chemical Bonding

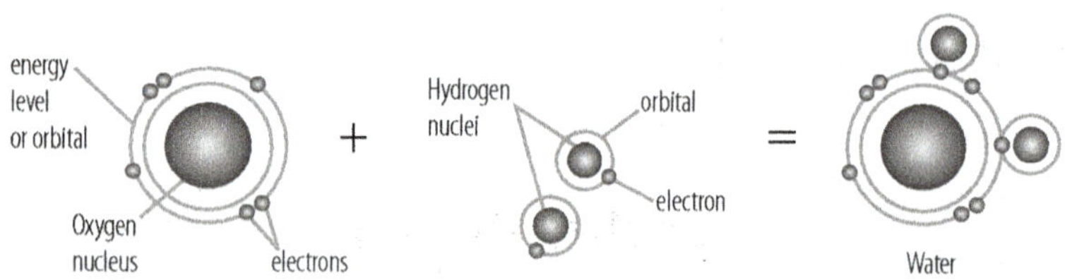

Sample Bonding Problems

CH_4	CO_2	C_2H_4	HCN	$C_6H_{12}O_6$

Biochemistry
Organic Molecules

- Carbohydrates
  - Made up of Carbon-Hydrogen-Oxygen
  - *Simple* (sugar)—found in fruit, candy, pop
    - Digested very rapidly (minutes)
  - *Complex* (starch)—found in bread, pasta, potatoes
    - Digested rapidly (within hours)
  - Used by the body as energy source

Carbohydrates

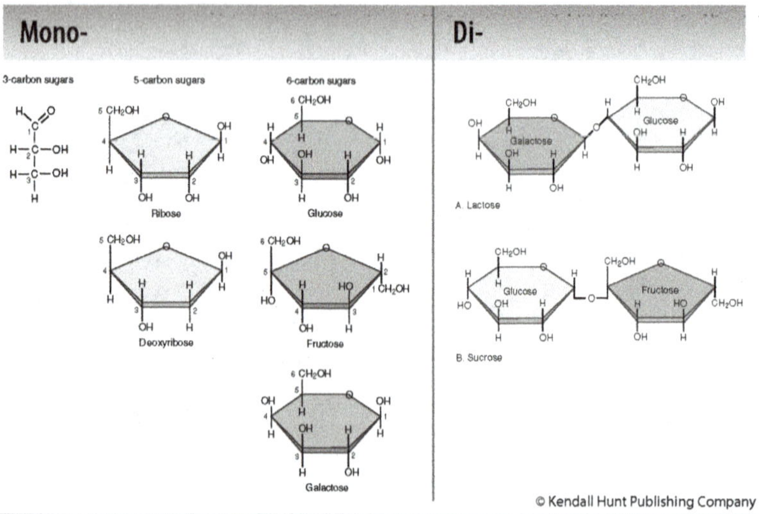

Biochemistry

Lipids

- Made up of a glycerol molecule and three fatty acids (Carbon-Hydrogen-Oxygen)
- Found in our diets as fats, oils, and waxes
- Digestion is very slow (several days to weeks)
- Used by the body for long-term energy storage and insulation

Lipids

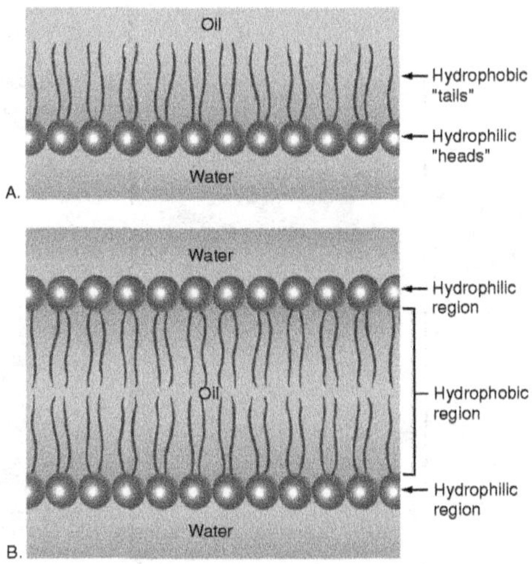

Biochemistry

Protein

- Large molecules made up of amino acids (C-H-O-N), the primary source of nitrogen
- Found in our diets as meat, beans, and nuts
- Digestion rate is slow (several hours to days)
- Used by the body in repair and maintenance
- Some proteins act as enzymes to speed up chemical reactions

Protein

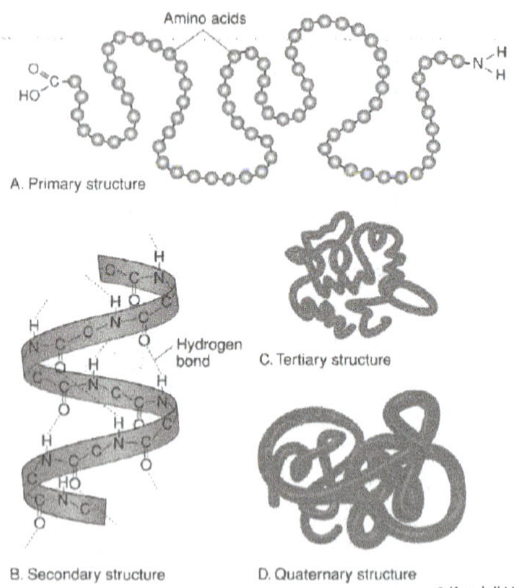

Biochemistry

Nucleic Acids

- Huge molecules made up of a sugar-phosphate-nitrogenous base
- Produced by the body as RNA and DNA
- Not digested by the body
- Used by the body as the "genetic code"

Nucleotide

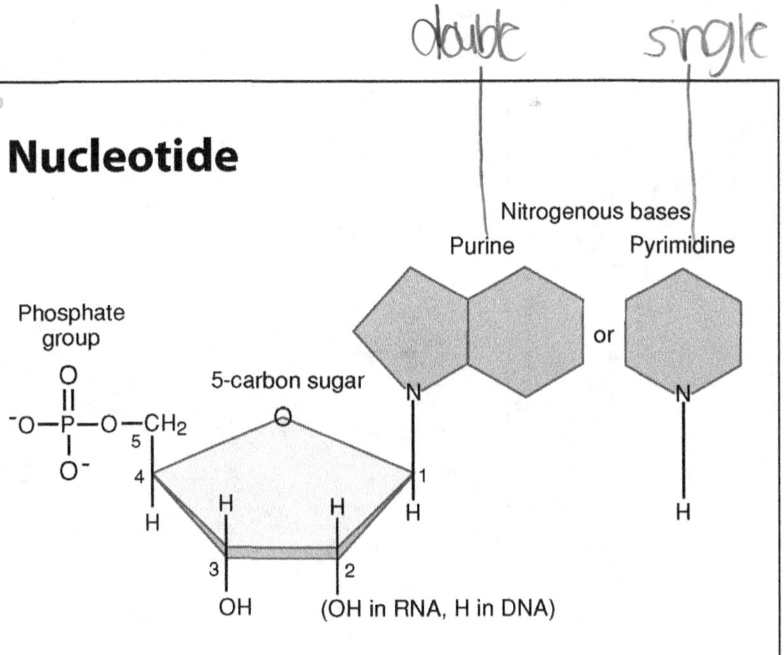

Nucleic Acids

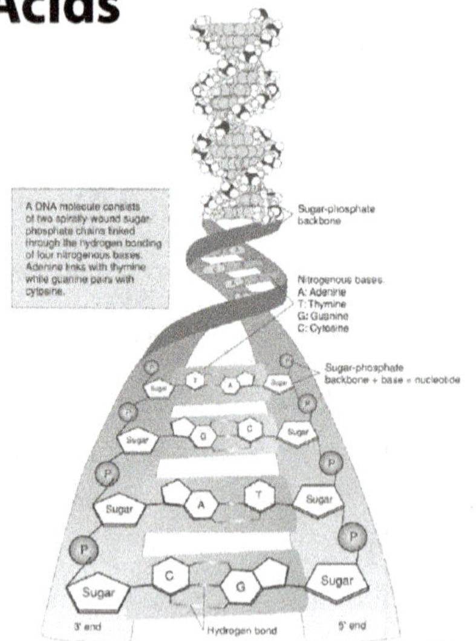

Nutrition

- Macronutrients
  - Carbohydrates
  - Protein
  - Fat
- Micronutrients
  - Vitamins
  - Minerals

Food Pyramid

Courtesy of USDA

Food Pyramid

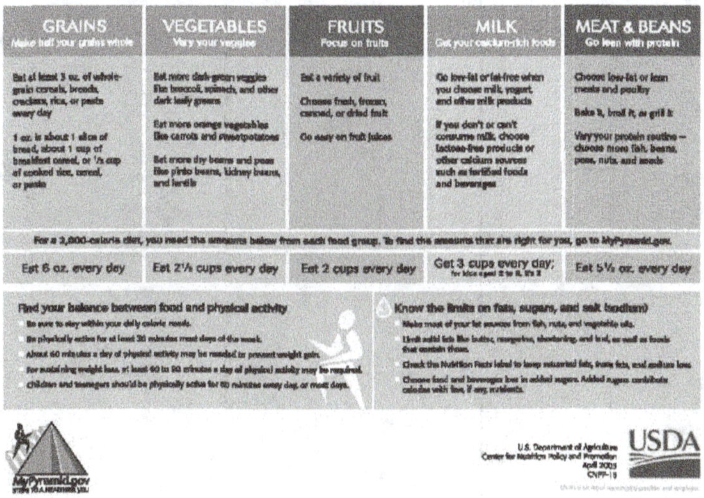

Courtesy of USDA

Carbohydrates

- Simple (sugar)—empty calories
- Complex (starch)
- Contain:
  - Carbon
  - Hydrogen
  - Oxygen
- Easily and rapidly digested
- Used for energy by the body

Glucose

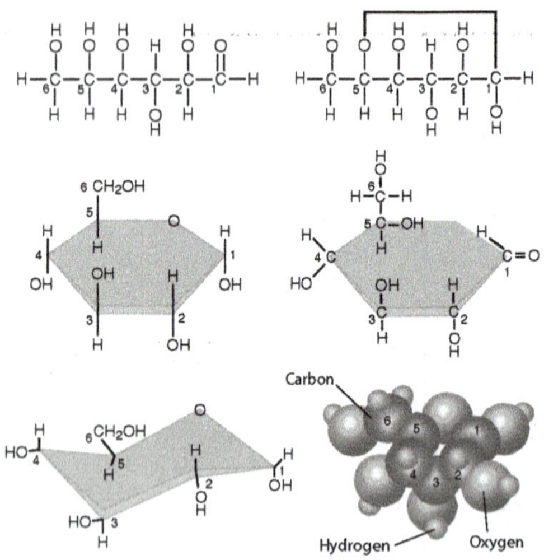

Carbohydrate Loading

- The protocol was originally developed in 1967 by *Swedish* scientist Gunvar Ahlborg
- If the body's glycogen stores were depleted, it would store more glycogen than normal when carbohydrate intake returned to normal

Carbo-Loading

- One week before the event:
  - three days of minimal carbohydrate intake (about 10% of total calories)
  - and intense exercise to deplete the body's carbohydrate stores
  - the next three days consume primarily carbohydrates (about 90% of total calories), and reduce the intensity of exercise to allow for maximum storage

Modified Carbo-Loading

- In the 1980s, research led to a modified carbo-loading regimen:
  - eliminates the depletion phase
  - instead calls for increased carbohydrate intake (to about 70% of total calories)
  - decreased training for three days prior to the event

Carbo-Loading

- The average male athlete can store (short term) about 1,500 to 1,900 kilocalories (kcal) of carbohydrate:
  - 60–80 kcal in the blood
  - 360–440 kcal in the liver
  - 1,300 to 1,400 kcal in the muscles

Carbo-Loading

- Two hours of exercise or a 20-mile run can deplete liver and muscle glycogen (stored carbohydrates) levels
- Obvious signs of *glycogen* depletion are heavy, tired muscles; poor performance; and possibly complete fatigue

Carbo-Loading

- Carbohydrates should make up at least 60% of the daily caloric intake of a non-athlete
- Less than that could mean that you may not be replenishing your liver and muscle glycogen stores

Carbo-Loading

- Eight to nine grams of carbohydrate per kilogram of body weight on a daily basis are needed
- One gram of carbohydrate equals 4 kilocalories and a kilogram equals 2.2 pounds

Proteins

- Contain:
    - Carbon
    - Hydrogen
    - Oxygen
    - Nitrogen
- Used for repair and maintenance of tissue
- Found in meats, dairy products, nuts, and beans
- Digested slowly (several hours)
- Provide the essential *Amino Acids*

Lipids

- Contain:
  - Carbon
  - Hydrogen
  - Oxygen
    - Much higher ratio of hydrogen to oxygen than 2/1
- Combination of one glycerol molecule and three fatty acids
- Found in fats, waxes, and oils
- Used by our body as long-term energy storage and insulation
- Very slow to digest (days)

Hawaiian Study

- Ancient diet from records of more than 1700 years ago
- The bulk of the diet of the original inhabitants of the islands consisted of plant material

Native Hawaiian Diet
Ancient

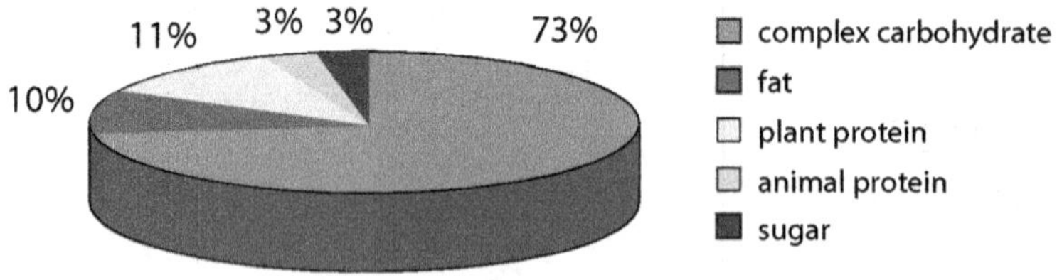

Native Hawaiian Diet
Modern

- Processed meats were introduced to the native inhabitants during World War II
- Cheap, plus easily stored and transported to the islands
- The modern diet depends on animal protein rather than plant protein and is 42% saturated fat

Native Hawaiian Diet
Modern

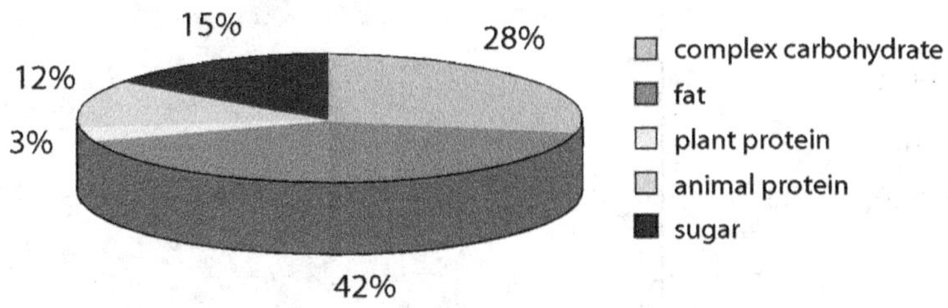

Hawaiian Study

- In a single generation
- Statistical studies showed—
  - among those native Hawaiians who have switched back to the native diet
  - the incidence of cardiovascular disease, cancer, obesity, and diabetes has dropped dramatically

Eating Disorders
Obesity

- **Obesity**—a weight 20% or more above the ideal weight for a person's height
- In U.S.—28% of women and 10% of men are obese
- **Moderate Obesity**—41–100% above ideal weight
- **Severe Obesity**—100% or more above ideal weight

Eating Disorders
Bulimia Nervosa

- Recurrent episodes of binge eating
- Characterized by consuming a much larger amount of food than normal for one sitting
- Combined with a sense of lack of control over eating during the episode
- An obsession about body weight and shape

Eating Disorders
Bulimia Nervosa

- Body weight is regulated by:
  - A restrictive diet
  - Excessive exercise
  - Purging (self-induced vomiting)
  - Laxative misuse/abuse

Eating Disorders
Anorexia Nervosa

- A morbid fear of gaining weight
- Body weight no more than 85% of normal
- A distorted body image so that person feels fat even when emaciated
- An absence of menstrual cycle for at least three months (in females)

Eating Disorders
Anorexia Nervosa

- Body weight is kept too low by either/or:
    - A restrictive diet often with excessive exercise
    - Binge eating/purging

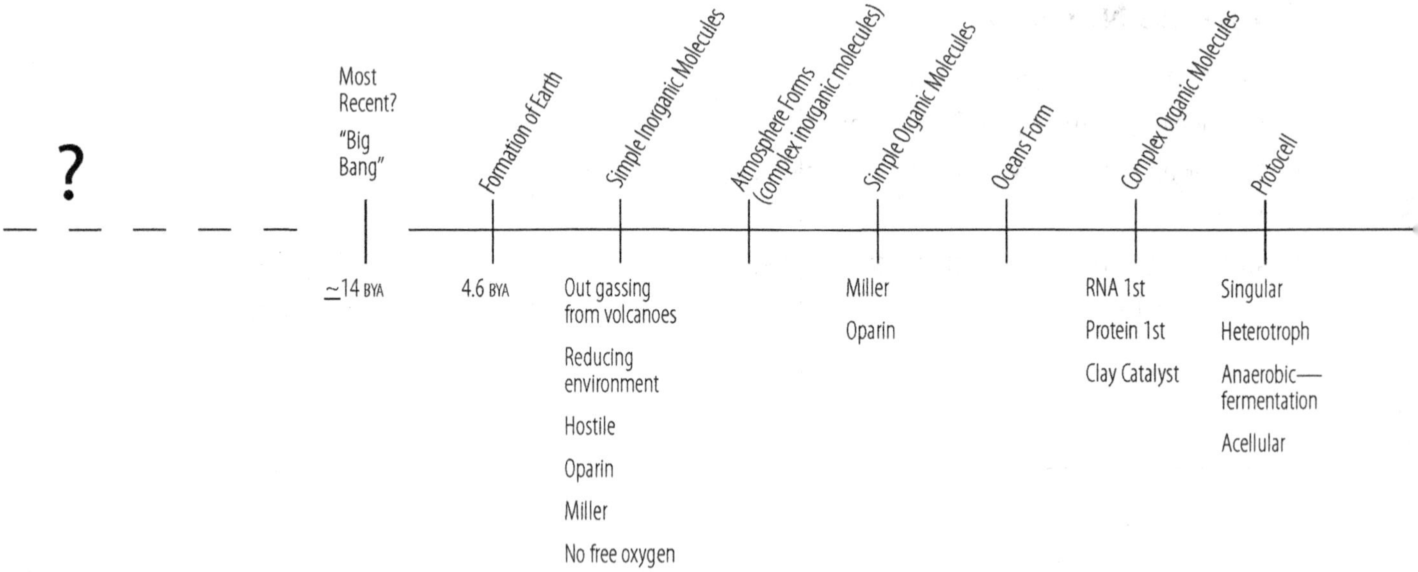

continued

Origin of Life

Extraterrestrial Origin

- Life did not originate on earth
- May have been carried to earth as an infection of spores
- Traveled via asteroid, comet, or meteorite
- How or when life developed here is not likely to be discovered soon

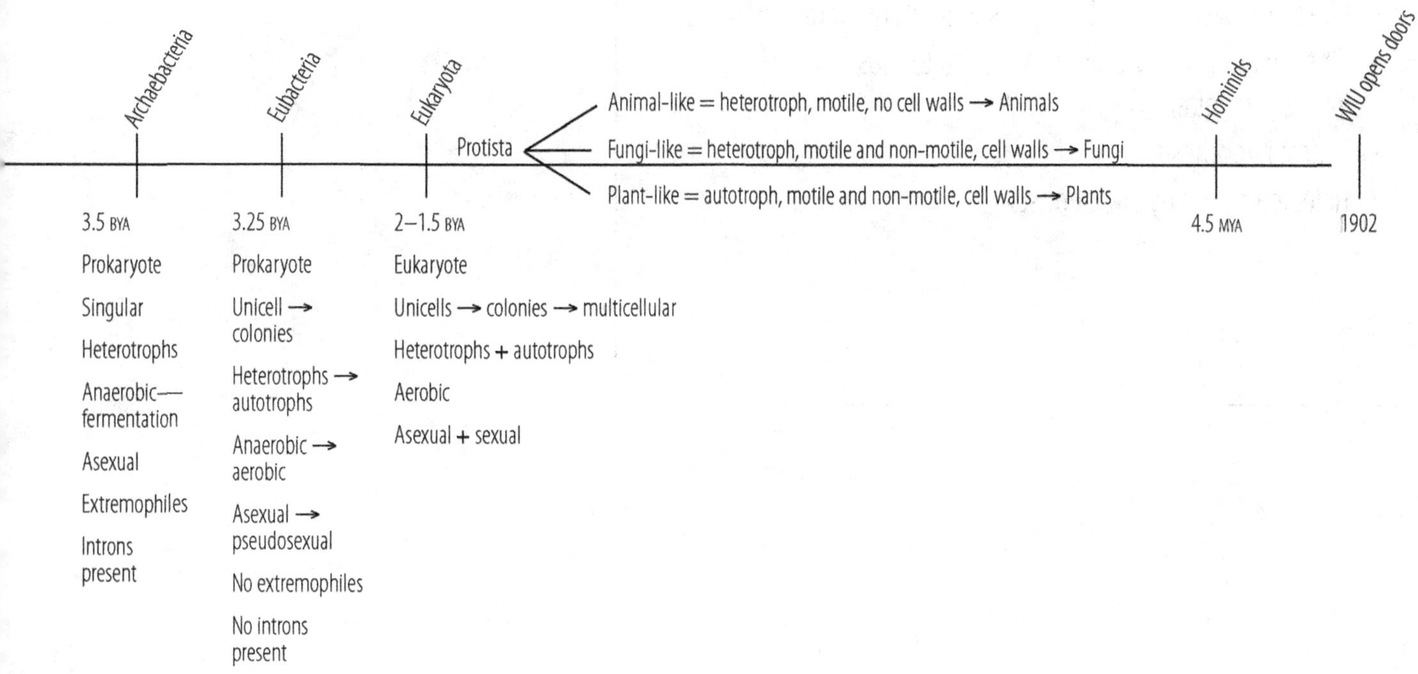

Extraterrestrial Origin

© NASA

Origin of Life
Special Creation

- Life forms were put on Earth by supernatural or divine forces
- This Creationist viewpoint is common to most organized Western religions
- Oldest hypothesis
- Rejected by nearly all scientists

Origin of Life

- Evolution
- Life may have evolved from inanimate matter
- With association among molecules becoming more and more complex
- The force leading to life was selection - natural
- Changes in molecules that increased their stability caused the molecules to persist longer

Origin and Evolution of Life

Primitive Atmosphere

- Produced by out gassing from the interior
- Primarily from volcanoes
- Contained no free oxygen
- Contained:
  - Water vapor, methane, ammonia, carbon dioxide, small amounts of hydrogen and carbon monoxide
- Was a reducing rather than oxidizing atmosphere of today

Primitive Atmosphere

Reducing vs. Oxidizing

- LEO says GER

L-Lose	E-Electron	O-Oxidize
G-Gain	E-Electron	R-Reduce

Oxidation Reduction Reactions

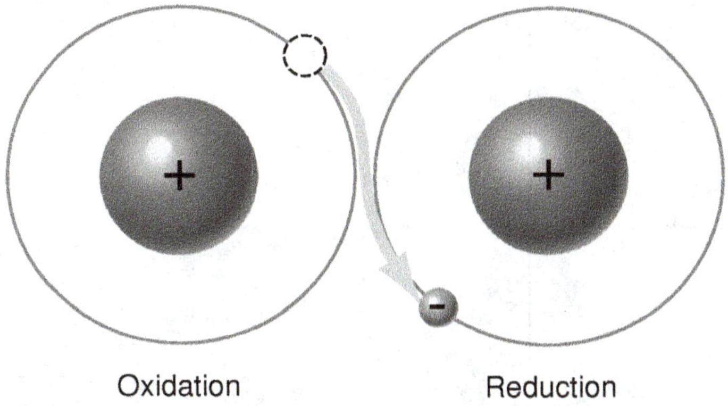

Stanley Miller's Experiment
Organic Molecules from Primitive Atmosphere

- Oparin (1938) hypothesized that the original organic molecules appeared naturally from chemicals in the atmosphere
- Miller (1953) experimented with a chamber filled with gases thought to be abundant in the primitive atmosphere
- Bombarding these with electrical charges (to simulate the input of energy through lightning)
- Amino acids and other organic molecules were formed

Miller's Experiment

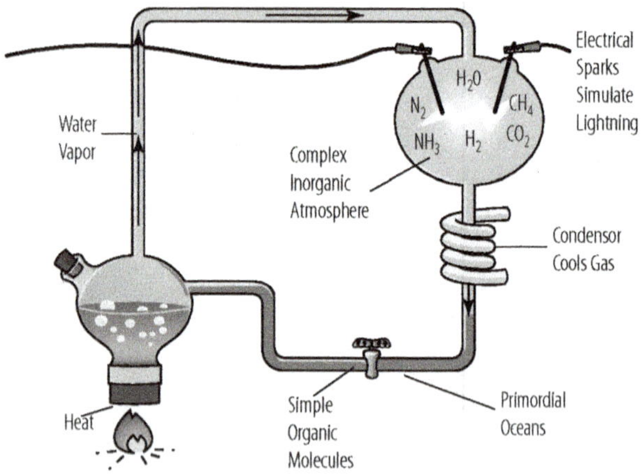

RNA–First Hypothesis

- Suggests that only RNA was needed for the series of steps that ended in the formation of the first cell
- Formulated after the discovery that RNA can act both as a substrate and an enzyme
- It could carry out the functions of both DNA and proteins

Protein–First Hypothesis

- Introduced by Sidney Fox
- Said amino acids polymerize when exposed to dry heat
- Suggests that amino acids collected in shallow puddles along the rocky shore and the heat from the sun caused them to become *proteinoids*
- **Proteinoids**—small polypeptides that had some catalytic properties
- Assumes DNA genes came after protein enzymes formed

Cairns-Smith Hypothesis

- Clay was particularly helpful in causing the polymerization of both proteins and nucleic acids at the same time
- Clay attracts small organic molecules
- Clay contains iron and zinc that may have served as catalysts for polypeptide formation
- Clay collects energy from radioactive decay and releases it when temperature and/or humidity change (energy source for polymerization)
- This suggests that polypeptides and RNA arose at the same time

The Protocell

- A structure that had a lipid-protein membrane and carried on energy metabolism
- Microspheres—composed only of protein, have a number of cellular characteristics
- Coacervate droplets—(Oparin's Hypothesis) have a tendency to absorb and incorporate various substances from the surrounding solution

The Heterotroph Hypothesis

- The protocell was most likely heterotrophic (takes in preformed food for energy)
- Lived in the primitive ocean where there was a ready supply of small organic molecules including ATP
- Probably had a limited ability to break down complex molecules for energy
- Since there was no free oxygen, it must have been a form of fermentation

The True Cell

- A membrane bound structure
- Can carry on protein synthesis that can produce the enzymes needed for DNA replication to take place
- Capable of reproducing itself
- DNA-RNA-Protein is the standard accepted relationship today

Prokaryotic Cell

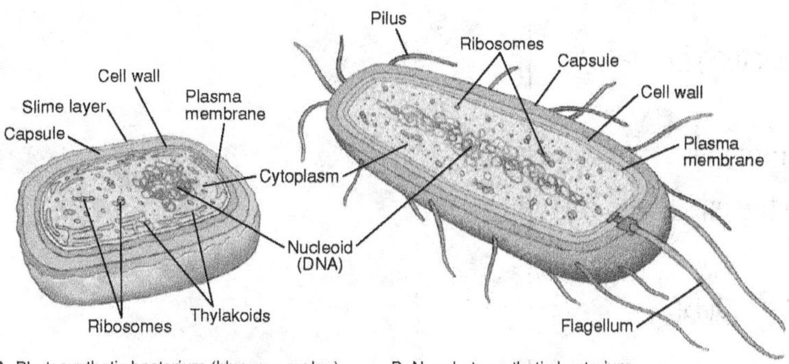

A. Photosynthetic bacterium (blue-green alga)   B. Nonphotosynthetic bacterium

Origin of Life
Evolution of Eukaryotic Cells
Endosymbiotic Hypothesis

- Prokaryotic cells enter a symbiotic relationship with other prokaryotic cells
- Small cells enter a large cell through endocytosis
- This creates a large cell with small specialized cells (organelles) inside
- When the large cell divides, so do the small cells
- This may have produced the first eukaryotic cell

Endosymbiotic Hypothesis
Evidence

1. Mitochondria and chloroplasts are similar to bacteria in size and structure
2. Both organelles are bounded by a double membrane:
   - The outer membrane may be derived from the engulfing vesicle
   - The inner membrane may be derived from the plasma membrane of the original prokaryote

Endosymbiotic Hypothesis
Evidence

3. Mitochondria and chloroplasts contain a limited amount of genetic material and divide by splitting
   - Their DNA is a loop like that of bacteria
4. Although most of their proteins are produced by the eukaryotic host:
   - They have their own ribosomes
   - They produce their own proteins
   - Their ribosomes resemble those of bacteria

Endosymbiotic Hypothesis
Evidence

5. The RNA (ribonucleic acid) base sequence of their ribosomes suggests a eubacteria origin for chloroplasts and mitochondria
6. Flagella may have been derived from an elongated bacterium that became attached to the host

Multicellularity

- Some unicellular organisms began living in association with others
- Safety in numbers
- Take advantage of special abilities of some organisms
- Unicellular → Colonies → Multicellular

Eukaryotic Cell

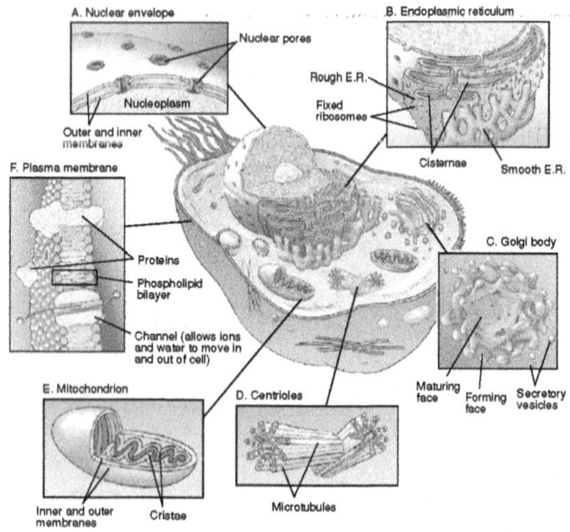

UNIT ONE WRAP-UP!

In preparation for this exam, I recommend the following steps:

1. Review your notes from class.
2. Make sure you translate each page from my language into yours.
3. Go to the Web links on metric conversions: Practice until you are comfortable with the procedure.
4. Review the periodic table until you understand why each of the elements are located where they are and what subatomic particles they are made of.
5. At the very last minute, use your copy of my text, written in your language to "teach" this material to your partner. Now, exchange roles and let her "teach" it to you. Ask questions until you are both satisfied with your grasp of the subject matter.

Helpful Exam Hints

- Be rested.
- Be focused.
- Be hydrated.
- Never work in groups larger than three (it becomes counterproductive).
- Relax (it will be fine—I promise).
- Don't wait until the last minute to take your exam.
- Never change an answer (unless upon checking your answers, you find you had misread the question).
- Make sure you have had a healthy meal within the last six hours.
- Read each question carefully and answer to the best of your ability.
- You may revisit all questions until you submit your exam.
- Your exam is timed so be aware of your time remaining; don't spend all of your time looking up the answer to one question.

Unit Two

Why doesn't mine look like hers?

DOES THIS CONTRACEPTION WORK EVERY TIME?

Did Al Capone really die from syphilis?

What are the thirteen steps to proper condom use?

Human development—or did we all really look like tadpoles once?

DOES THIS LOOK RIGHT TO YOU?

Sexual Reproduction

EXAM II

Human Reproduction
Anatomy and Physiology
Contraception
Sexually Transmitted Disease
Human Development

Reproduction

Male Reproductive System

- **Vas Deferens**—tube connecting epididymis to urethra
- **Epididymis**—location of sperm maturation
- **Ejaculatory Duct**—location of the storage of mature sperm
- **Urethra**—tube that carries sperm and urine from the body
- **Penis**—male organ of copulation
- **Glans Penis**—enlarged tip of penis
- **Foreskin**—thin layer of skin that covers the glans penis

Male Reproductive System

- **Testis**—sperm and sex hormone production
- **Seminal Vesicle**—adds secretion to ejaculate
- **Bulbourethral Gland**—adds secretion to ejaculate
- **Prostate Gland**—adds secretion to ejaculate
- **Scrotum**—sac that holds testis and regulates the temperature of the sperm

Male Reproductive System

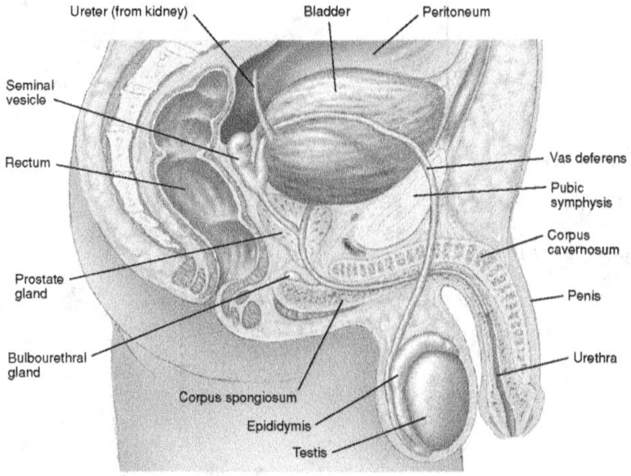

Male Reproductive System

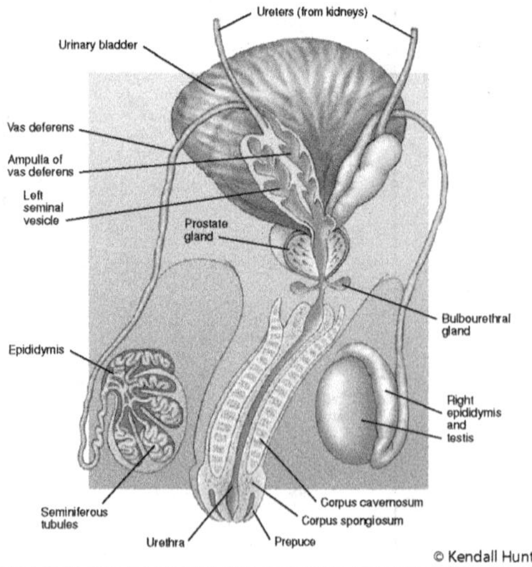

Reproduction
Sperm

- **Sperm**—made up of three parts:
  1. *Head*—Carries genetic information, covered by the Acrosome: produces an enzyme that digests the wall of the egg
  2. *Middle Piece*—contains mitochondria that supply energy to the sperm
  3. *Tail*—allows for the motility of the sperm

Sperm

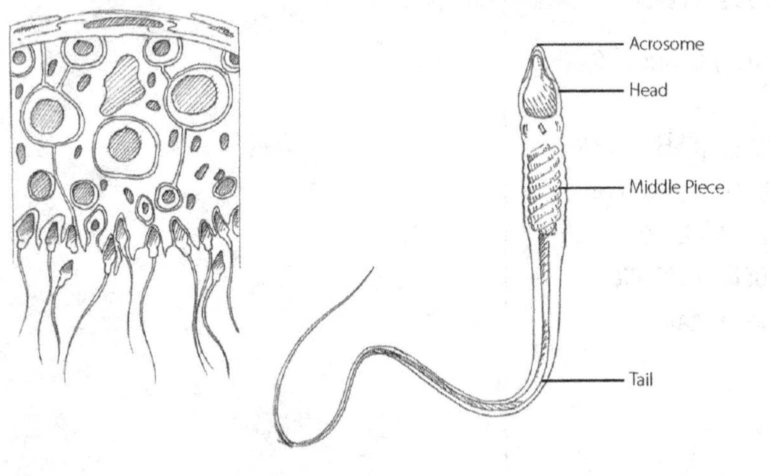

Image courtesy of Kevin Mercer

Spermatogenesis

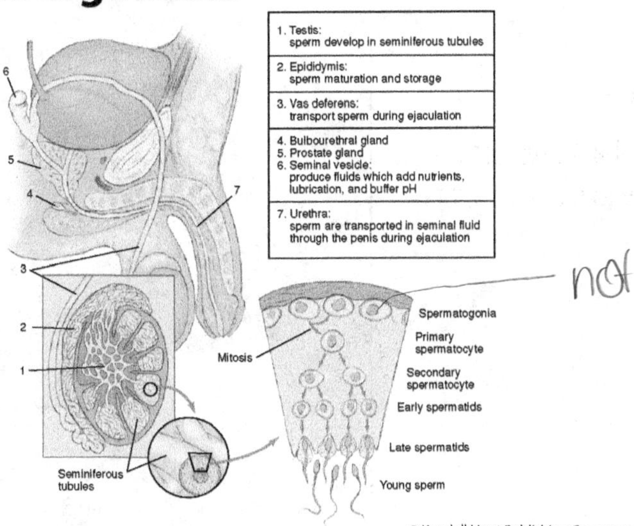

Hormonal Regulation in Males

- **Hypothalamus**—secretes the hormone (GnRH) *gonadotropin-releasing hormone* this stimulates . . .
  - *Anterior Pituitary*—to secrete two hormones: (FSH) *follicle stimulating hormone*, which promotes the production of sperm and *inhibin* in the testis and (LH) *luteininzing hormone* which controls the production of *testosterone*
    - *Testosterone*—is the main sex hormone in males

Reproduction
Hormonal Regulation in Males

- Testosterone and inhibin exert negative feedback pressure on the hypothalamus and the anterior pituitary to maintain proper levels of testosterone in the blood

Female Reproductive System

- **Ovaries**—produce egg and sex hormone
- **Oviducts**—conduct egg, location of fertilization (Fallopian tubes)
- **Uterus** (Womb)—houses developing fetus
- **Cervix**—contains opening to uterus
- **Vagina**—female organ of copulation, serves as birth canal, and as exit of menstrual fluids

Female Reproductive System

- **Endometrium**—lining of the uterus, participates in the formation of the placenta
- **Placenta**—supplies nutrients needed for embryonic and fetal development
- **Vulva**—collective term for external genitalia of the female
- **Labia Majora**—two large, hair-covered folds of skin

Female Reproductive System

- **Labia Minora**—two small folds of skin just inside the labia majora
- **Clitoris**—an organ that is homologous to the penis, has a shaft of erectable tissue
- **Glands Clitoris**—tiny cap of sense receptors covering the clitoris
- **Hymen**—membrane that partially blocks the opening of the vagina

Female Reproductive System

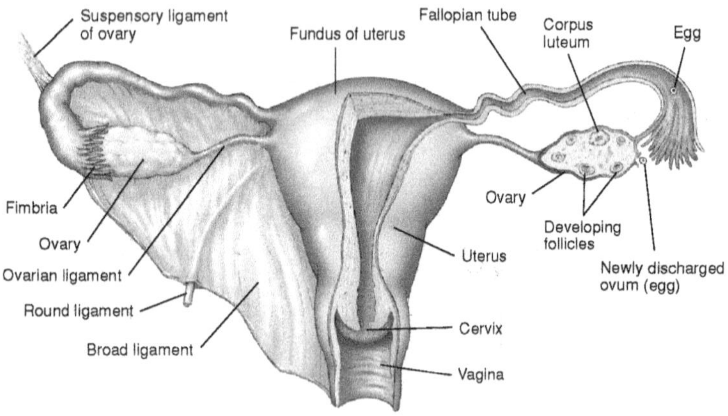

© Kendall Hunt Publishing Company

Ovarian Cycle

- **Follicular Phase**—(days 1–13: first half of the cycle)
  - *Hypothalamus*—produces (GnRH)
  - *Anterior Pituitary*—produces *FSH* and *LH* in response to the increased levels of *GnRH*
  - *FSH* promotes the development of a follicle, which secretes *estrogen*, as estrogen levels increase, this exerts a feedback control to the anterior pituitary, stopping the production of *FSH* and bring the Follicular phase to an end

Ovarian Cycle

- **Ovulation**—(day 14) high levels of *estrogen* in the blood causes the Hypothalamus to secrete high levels of *GnRH*
  - *Anterior Pituitary*—produces a surge of *LH* which stimulates ovulation
  - Release of a secondary oocyte from the follicle of the ovary

Ovarian Cycle

- **Luteal Phase**—(days 15–28: last half of the cycle)
  - *LH*—promotes the development of the corpus luteum, which secretes progesterone
  - *Progesterone*—causes the uterine lining (endometrium) to build up, as levels rise it causes a feedback control over the anterior pituitary to stop production of LH, causing the corpus luteum to degenerate, bringing about menstruation (the end of the luteal phase)

Ovarian Cycle

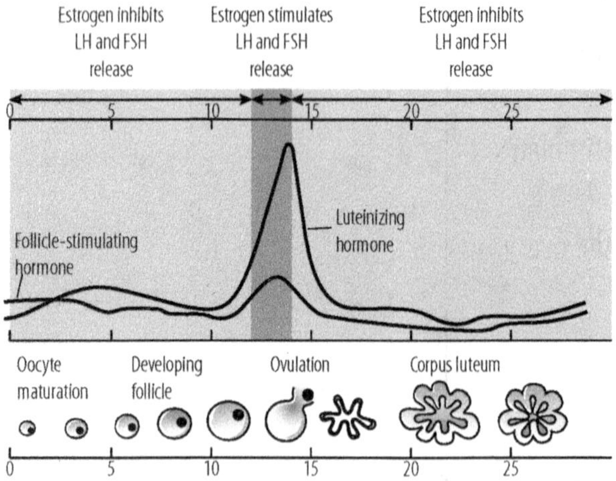

Oogenesis

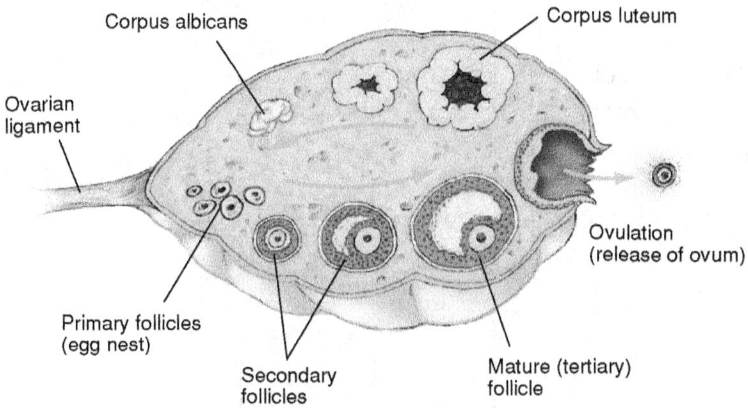

Uterine Cycle

- **Menstruation** (days 1–5)—low levels of *estrogen* and *progesterone* cause the endometrial lining to disintegrate and its blood vessels to rupture, a flow of blood and tissues (menses) passes out of the vagina
- **Proliferative Phase** (days 6–13)—increased levels of *estrogen* produced by a new follicle cause the endometrium to thicken and become vascular and glandular

Uterine Cycle

- **Ovulation**—occurs on the 14th day of the cycle
- **Secretory Phase** (days 15–28)—increased production of progesterone by the corpus luteum causes the endometrium to double in thickness and the uterine glands to mature

The uterus is now ready for the implantation of the developing embryo

Uterine Cycle

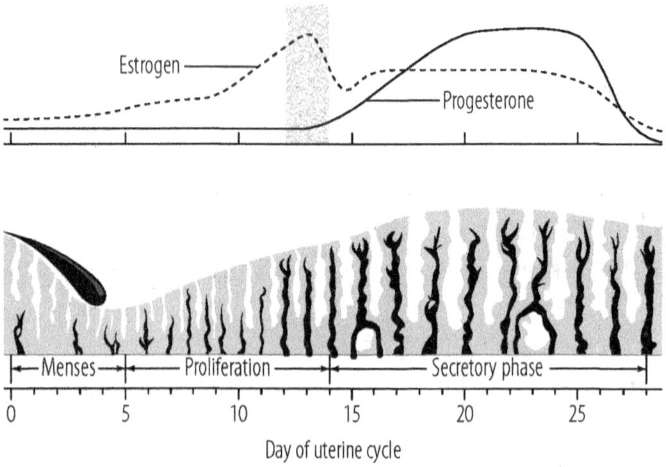

Reproduction

Fertilization and Pregnancy

- **Fertilization**—union of sperm and egg nuclei, restoring the diploid number of chromosomes to the cell
- **Placenta**—originates from both maternal & fetal tissue, it is the site of exchange of molecules between maternal and fetal blood

There is no menstruation during pregnancy

Contraception

- **Abstinence**—refrain from sexual intercourse, no sperm in vagina, 100% effective no risks

Surgical

- **Vasectomy**—*Vas deferentia* are cut and tied, no sperm in seminal fluid, almost 100% effective, can cause irreversible sterility
- **Tubal Ligation**—oviducts are cut and tied, no eggs in oviduct, almost 100% effective, can cause irreversible sterility

Vasectomy

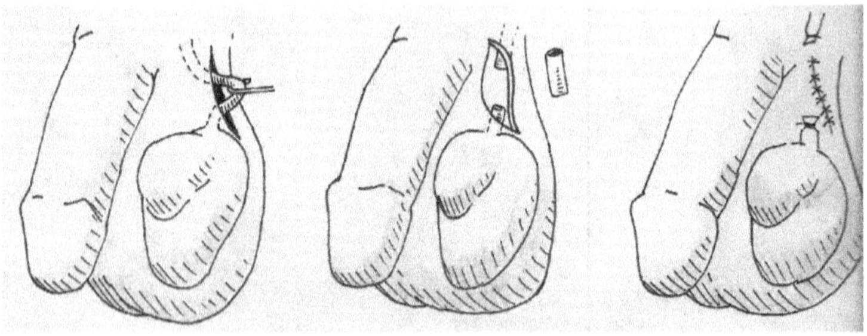

Image courtesy of Kevin Mercer

Tubal Ligation

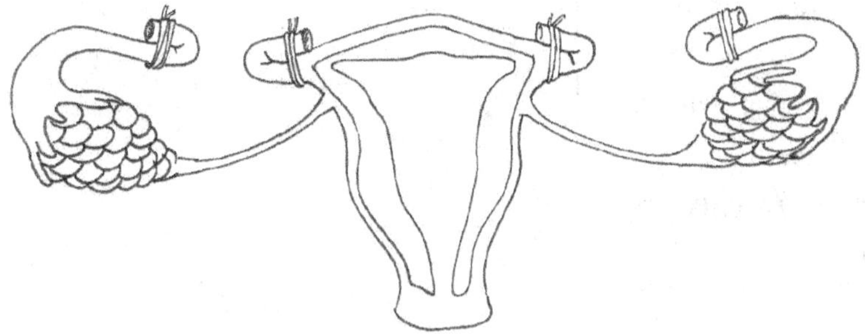

Image courtesy of Kevin Mercer

Hormone Replacement

- **Oral Contraceptive**—hormone medication is taken daily, anterior pituitary does not release FSH and LH, almost 100% effective, can cause Thromboembolism (blood clotting)
- **Depo-Provera Injections**—4 injections of progesterone-like steroid are given per year, anterior pituitary does not release FSH and LH, about 99% effective, can cause blood clotting or may lead to breast cancer or osteoporosis

Hormone Replacement

- **Orthoevra**—(Patch) epidermal hormone medication, anterior pituitary does not release FSH and LH, almost 100% effective, can cause thromboembolism (blood clotting)
- **Emergency Contraception**—hormone medication is taken in the event of an unplanned sexual experience. Powerful oral contraceptive administered afterwards, almost 100% effective, can cause thromboembolism (blood clotting)

Hormone Replacement

- **Contraceptive Implants**—tubes of progestin are implanted under the skin, anterior pituitary does not release FSH and LH, >90% effective, can cause blood clotting or may lead to breast cancer or osteoporosis
- **Intrauterine Device (IUD)**—plastic coil is inserted into the uterus by a physician, prevents implantation, >90% effective, can cause infection (Most used form of contraceptive World Wide)

Barrier Methods

- **Diaphragm**—latex cup inserted into the vagina to cover cervix before intercourse, block entrance of sperm to uterus, with jelly about 90% effective, presently no known risks
- **Cervical Cap**—latex cup is held by suction over cervix, delivers spermicide near cervix, almost 85% effective, may cause cancer of the cervix

Barrier Methods

- **Male Condom**—latex sheath is fitted over erect penis, traps sperm and prevents many STDs, about 85% effective, presently no known risks
- **Reality Condom**—polyurethane tubing is fitted inside vagina, blocks entrance of sperm to uterus and prevents many STDs, about 85% effective, presently no known risks

Contraception

- **Coitus Interruptus**—male withdraws penis before ejaculation, prevents sperm from entering vagina, about 35% effective, presently no known risks
- **Jellies, Creams, and Foams**—spermicidal products inserted before intercourse, kills a large number of sperm, about 75% effective, presently no known risks

Contraception

- **Natural Family Planning**—day of ovulation is determined by record keeping (various methods of testing), intercourse is avoided on certain days of the month, about 70% effective, presently no known risks
- **Douche**—vagina is cleansed after intercourse, washes sperm out, <25% effective, presently no known risks

Contraception

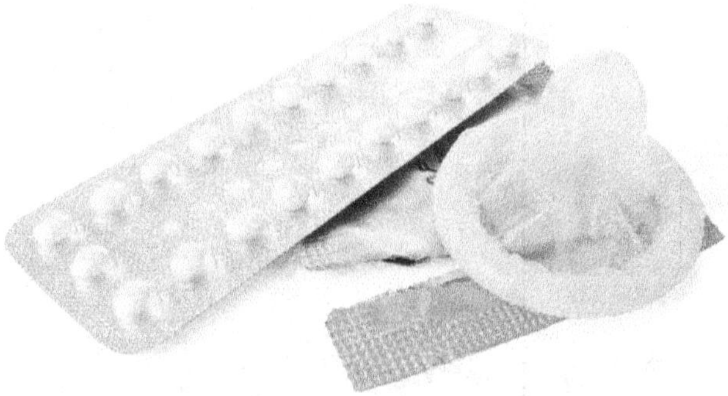

Image © ZTS, 2009. Used under license from Shutterstock, Inc.

Reproduction

Sexually Transmitted Diseases (STDs)

- **Acquired Immunodeficiency Syndrome (AIDS)**—caused by a related retrovirus known as HIV (human immunodeficiency viruses) that infect specific blood cells
  - >40 million people have tested positive for HIV
  - >12 million people have died of AIDS
  - A new infection occurs every 15 seconds

The majority of the new infections are among heterosexuals

Reproduction
Sexually Transmitted Diseases (STDs)

- Transmission
  - Sexual contact with an infected person
  - Needle sharing with an infected person
  - Babies born to infected mothers may become infected before or during birth, or through breast feeding

Reproduction
Sexually Transmitted Diseases (STDs)

- Three Stages of HIV:
  - *Category A*—helper T lymphocytes count is 500/mm cubed or greater, there maybe no symptoms, the person is highly infectious, blood test will still be negative for HIV
  - *Category B*—several months to several years after infection, helper T lymphocytes is 200–499/mm cubed, swollen nodes and fatigue are common symptoms

Reproduction
Sexually Transmitted Diseases (STDs)

- Three Stages of HIV *(continued)*:
  - *Category C*—helper T cell count is down to only 200/mm cubed, lymph nodes degenerate, severe weight loss, due to persistent diarrhea this weakened state of the immune system is what allows opportunistic infections a foothold

Reproduction
Sexually Transmitted Diseases (STDs)

- Opportunistic Infections
  - *Pneumocystis carinii pneumonia*—the lungs become useless as they fill with fluid and debris
  - *Mycobacterium tuberculosis*—bacterial infection usually of the lungs, commonly found in lymph nodes and other organs in AIDS patients (becoming multi-drug resistant in general populace)

Reproduction

Sexually Transmitted Diseases (STDs)

- **Karposi's Sarcoma**—an unusual cancer of the blood vessels
- **Invasive Cervical Cancer**—cancer of the cervix
- ***Toxoplasim encephalitis***—single cell parasite common to cats, leads to loss of brain cells, seizures, and loss of sensation on one side the the body

Reproduction

Sexually Transmitted Diseases (STDs)

- **Hepatitis B**—about 50% of the persons infected have flu-like symptoms; can be acute lasting for 3–4 weeks, or it can be chronic that leads to liver failure and possibly death, transmitted in the same manner as AIDS

Reproduction

Sexually Transmitted Diseases (STDs)

- **Herpes**—caused by the herpes simplex virus
    - *Type 1*—causes cold sores and fever blisters
    - *Type 2*—causes genital herpes, leading STD today, after infection there are no symptoms, 2–20 days later, itching or tingling at the infected site, blisters become painful ulcerations, then clear-up while dormant; babies born to infected mothers my have brain damage, go blind or even die

Reproduction

Sexually Transmitted Diseases (STDs)

- **Genital Warts**—caused by human papillomaviruses (HPVs), may appear as warts or flat lesions on infected area, associated with cancer of the cervix (90–95% of all cases, most common in young teenagers who have had multiple sex partners)

Reproduction
Sexually Transmitted Diseases (STDs)

- **Gonorrhea**—caused by the bacterium *Neisseria gonorrehoeae*, male experiences pain on urination and has a thick, greenish-yellow discharge 3–5 days after contact with an infected partner, in the female the bacterium spreads to the oviducts causing PIDs (pelvic inflammatory disease), as a result the female will become sterile

Reproduction
Sexually Transmitted Diseases (STDs)

- **Chlamydia**—caused by the bacterium, *Chlamydia trachomatis*, more numerous than any other STD infections, 8–21 days after infection, males experience mild burning sensation, on urination, females have a vaginal discharge along with symptoms of urinary tract infection, if not treated it can cause PID or sterility

Reproduction

Sexually Transmitted Diseases (STDs)

- **Syphilis**—caused by the bacterium *Treponema pallidum*, penicillin has been used as an effective antibiotic
  - *Primary Stage*—a hard chancre indicates the site of infection
  - *Secondary Stage*—victim breaks out in a rash that does not itch even on the sole of the feet and palms of the hands
  - *Tertiary Stage*—lasts until the patient dies

Development

Early Developmental Stages

- Fertilization
  - Results in a zygote
  - Requires the sperm and the egg to interact
  - Restores the diploid number of chromosomes to the cell

Development
Early Developmental Stages

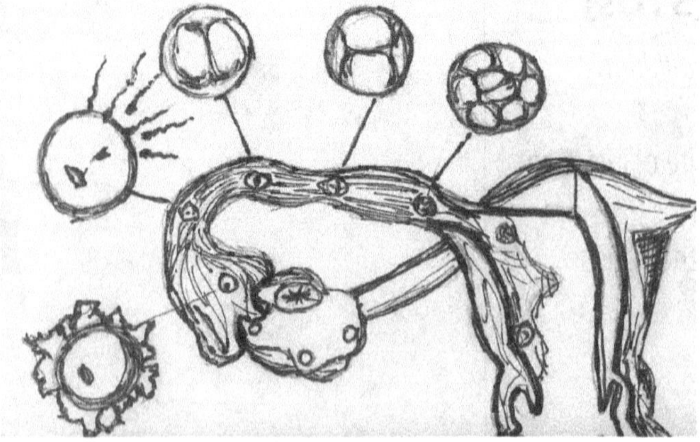

Image courtesy of Rebecca McGuire

Fertilization

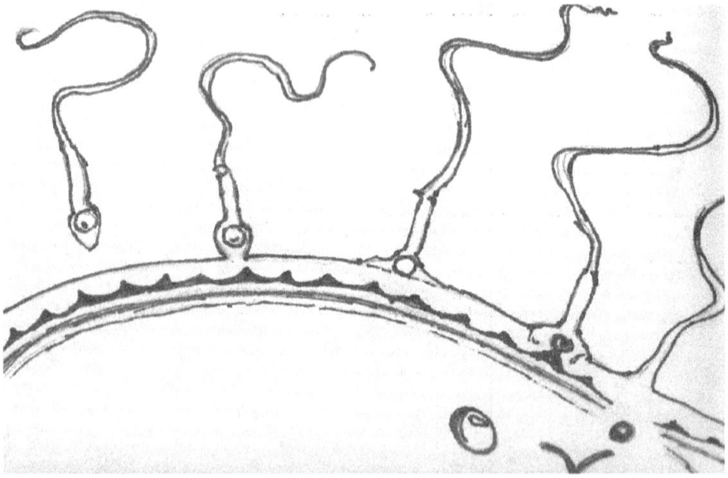

Image courtesy of Rebecca McGuire

Implantation

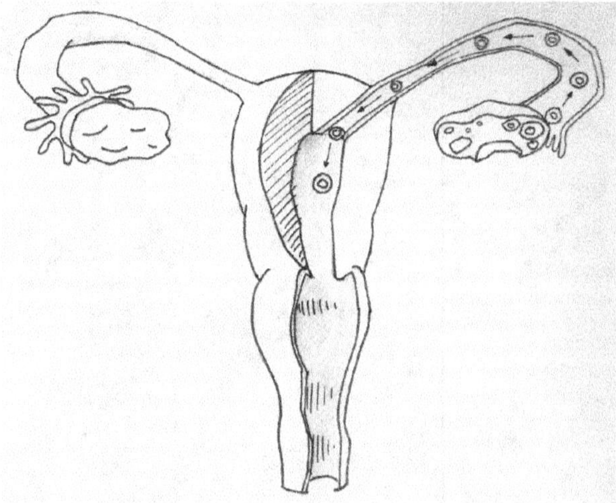

Image courtesy of Kevin Mercer

Early Developmental Stages

Cleavage

- 30 hrs.—3rd day
- Cell division with out growth
- DNA replication and cell division occur repeatedly
- Cells get smaller with each division
- Cell divisions are equal resulting in cells of equal size

Early Development
Cleavage

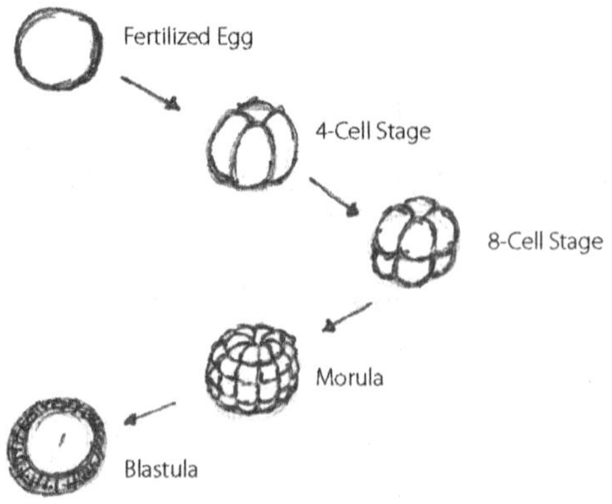

Image courtesy of Rebecca McGuire

Early Developmental Stages
Cleavage

- The four cell stage is the last point when all cells are identical (this is the last opportunity to clone the individual)
- *Morula*—solid ball of 16 cells (3rd–4th day)
- *Blastula*—hollow ball of cells (5th day–2nd week)
- *Blastocoel* cavity within the blastula (last stage of cleavage)

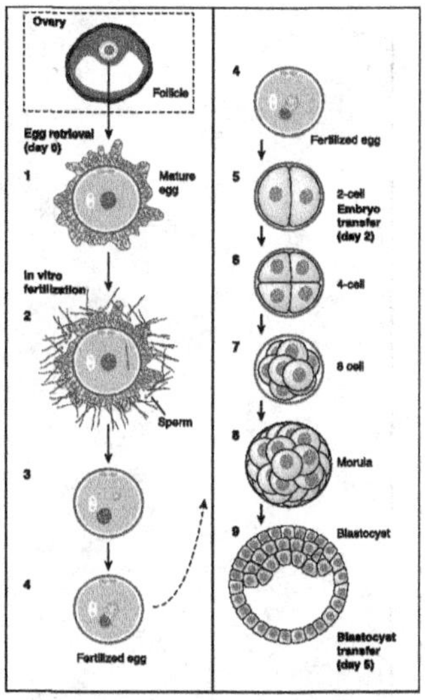

Early Developmental Stages

Gastrulation

- The *Gastrula* stage (end of 2nd week) begins when certain cells from the *Vegetal Pole* begin to push or invaginate into the blastocoel
- This creates a double layer of cells
  - *Ectoderm*—the outer layer of cells
  - *Endoderm*—the inner layer of cells
  - *Archenteron*—the space created by the invagination, will become the gut

Early Developmental Stages
Gastrulation

- This phase is not complete until all three embryonic germ layers of the cell are present
- *Mesoderm* middle layer of the cell
  - Begins as out pocketings of the Archenteron
  - These grow in size until they meet and fuse
  - Two layers of mesoderm are formed
- *Coelom* space between the two mesoderm layers

Gastrulation

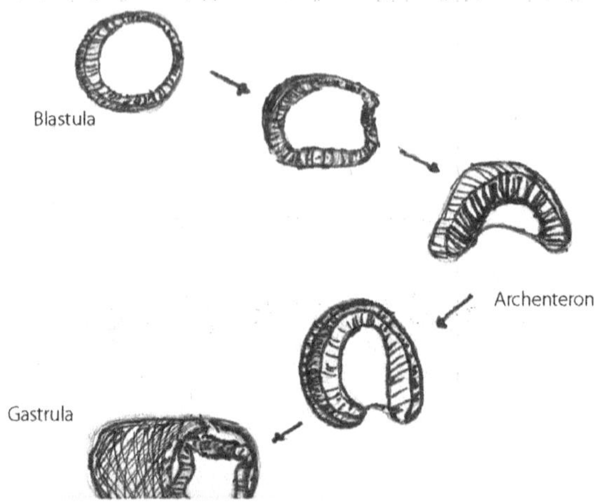

Image courtesy of Rebecca McGuire

Early Developmental Stages
Neurulation

- Mesoderm cells that lie along the longitudinal axis form a dorsal supporting rod called the *Notochord*
- This is later replaced by the *Vertebral Column* (in vertebrates) nervous system develops from the midline ectoderm located just above the notochord

Early Developmental Stages
Neurulation

- *Neural Plate* (thickening of cells)
- *Neural Folds* (develop on either side of the neural groove)
- *Neural Tube*—neural folds fused together
- *Neurula*—name of embryo at this point of development
  - Later the anterior end of the neural tube develops into the brain

Early Developmental Stages
Differentiation

- Occurs when ***cells*** become specialized in form and function
- Starts long before we can recognize different types of cells
- It is hypothesized that genes are turned on/off due to *Ooplasmic Segregation*
  - The distribution of maternal cytoplasmic contents to various cells of the morula

Early Developmental Stages
Morphogenesis

- Change in shape and form of a **body part**
- Significantly influenced by neighboring cells as well as cytoplasmic content
- **Induction**—process by which one tissue influences the development of another through contact or chemical signals

Early Developmental Stages
Human

- **Zygote**—from the point of fertilization to implantation
- **Embryo**—(1st and 2nd months) consists of early formation of organs
- **Fetus**—(3rd–9th months) consists of refinement of these structures

Early Developmental Stages
The First Week

- Fertilization occurs in the upper 1/3 of an oviduct
- Cleavage begins even as the embryo passes down the tube to the uterus
- *Blastocyst*—has a fluid filled cavity, a single layer of cells called the *Trophoblast* (food) and an *Inner Cell Mass* (which becomes the embryo and develops in to the fetus)

Early Developmental Stages

The Second Week

- Zygote begins the process of implanting in the wall of the uterus, now called an embryo
- Inner cell mass detaches itself from the trophoblast
- Gastrulation occurs

Early Developmental Stages

Embryonic Development

The Third Week

- The nervous system is the first to be visually evident
- Development of the heart begins and continues into the fourth week

Early Developmental Stages
Embryonic Development
The Fourth and Fifth Weeks

- The embryo is only 5–10 mm long
- The umbilical cord is fully formed
- Little flippers called limb buds appear
- The head grows and the sense organs become more apparent (Eyes, ears, and nose)

Early Developmental Stages
Embryonic Development
The Sixth through Eighth Weeks

- Form becomes easily recognizable as human
- Neck region develops
- Reflex action begin
- Embryo is 38 mm long and weighs about the same as an aspirin tablet
- All organ systems are now functional

Early Developmental Stages

Fetal Development

The Third Month

- The head is still very large
- The nose is flat
- The eyes are far apart
- The ears are distinctly present
- Eyebrows, eyelashes, hair on head, fingernails and nipples appear

Early Developmental Stages

Fetal Development

The Third Month

- Ossification begins (Cartilage is replaced by bone)
- Sometimes possible to distinguish between male and female

Early Developmental Stages
Fetal Development
The Fourth Month

- The fetal heart is loud enough to be heard by a stethoscope
- At the end of the month the fetus is about 152 mm (6 in) in length and about 171 g (6 oz)

Early Developmental Stages
Fetal Development
The Fifth through Seventh Months

- The mother begins to feel movement
- Covered in a fine down called *Lanugo*
- *Vernix caseosa*—white, greasy, cheese like substance that covers the fetus and protects it from the amniotic fluid

Early Developmental Stages

Fetal Development

The Fifth through Seventh Months

- At the end of this period the fetus has grown to about 300 mm (12 in) and 1,380 g (3 lbs)
- Babies born at this point have a good chance of surviving

Birth

- At the end of the 9 month gestation
- The baby is now about 20½ in and 7½ lbs
- The fetus rotates so the head is pointing toward the cervix
- The onset of labor is marked by uterine contractions that occur regularly every 15–20 minutes and that last for about 40 seconds or more

Parturition
Labor and the Expulsion of the Fetus

- First Stage:
  - Dilation of the cervix
  - Expulsion of the mucus plug from the cervical canal
- **Effacement**—the cervical canal slowly disappears as the head pushes on the lower part of the uterus

Parturition
First Stage—Cervix Dilates

- The head acts as a wedge to further dilate the cervix
- Since the head has a diameter of about 10 cm, the cervix must dilate to that diameter to allow the head to pass through
- If it has not happened already, the amniotic membrane will now rupture
- This stage comes to an end when the cervix is completely dilated

Parturition
Second Stage—Infant Emerges

- Uterine contractions occur about every 1–2 minutes and last about a minute each
- As the head descends into the vagina the desire to push increases
- The head turns so that the back of the head is uppermost when it appears (Internal Rotation)
  - **Episiotomy**—incision which enlarges the vaginal opening to prevent tearing

Parturition
Second Stage—Infant Emerges

- As soon as the head delivers the shoulders rotate so that the face is either to the right or left (External Rotation)
- Once the infant is breathing normally the ***Umbilical Cord*** is cut and tied
- The stump of the cord shrivels and leaves a scar (Belly Button or Navel)

Parturition

Third Stage—Expelling the Afterbirth

- About 15 minutes after delivery of the infant the uterine contractions dislodge the placenta
- This separates the placenta from the mothers blood supply and prevents hemorrhaging
- The placenta is expelled into the vagina
- This stage is complete when the placenta and its membranes have been delivered

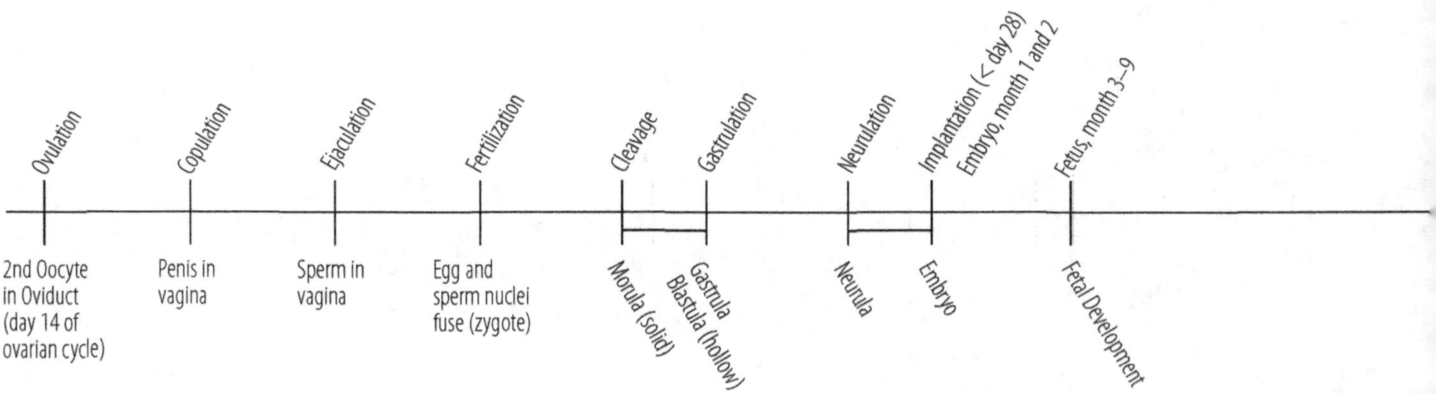

(continued)

Breast Feeding

- Suckling stimulates nerve endings in the nipple and areola of the breast
- Neural pathways now stimulate the Hypothalamus to produce Oxytocin
- Oxytocin travels to the Posterior Lobe of the Pituitary where it is released
- Causing mammary lobules to contract forcing out the milk
- Natural drop in Estrogen levels after birth stimulate the release of Prolactin from the Anterior Pituitary
- Continued stimulation maintains high levels of Prolactin which stimulate the production of milk

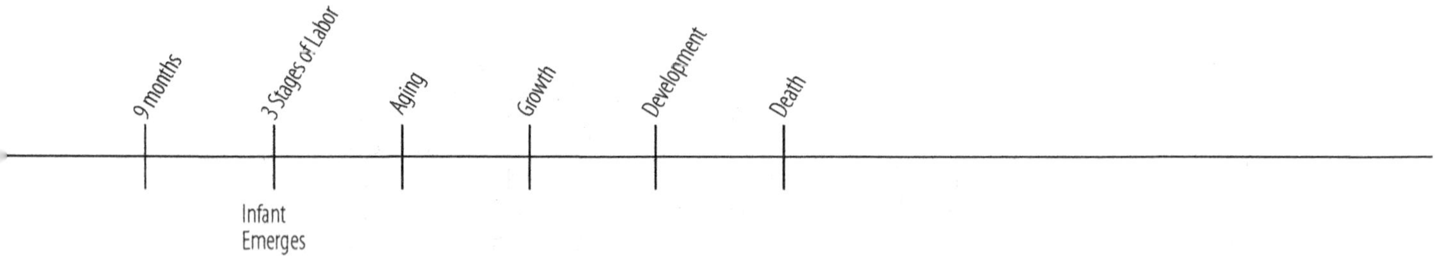

Aging Theories
Genetic

- The number of times a cell divides is species specific
- Some cell lines become dysfunctional before others
- The children of long lived parents typically live longer

Aging Theories
Whole–Body Process

- Decline in hormonal system can affect different organs in the body
- Overall decline in the functioning of the immune system
- Possible change in tissue that affect all organs

Effects of Aging on Body Systems
Skin

- **Less Elastic**—due to a decrease in elastic fibers and an increase of collagen fibers
- **Less Adipose Tissue**—decreasing the insulative property of skin
- **Fewer Sweat Glands**—causing a decrease in homeostatic adjustment due to decreased ability to sweat

Effects of Aging on Body Systems
Skin

- **Fewer Hair Follicles**—the hair thins out in both sexes
- **Fewer Oil Glands**—skin tends to dry and crack

diet, alchol & tabacco, exerase, sun

Hormone	Origin	Target	Tissue	Response
Gonadotropin-Releasing Hormone				
Follicle Stimulating Hormone				
Luteininzing Hormone				
Progesterone				
Estrogen				
Testosterone				
Inhibin				
Human Growth Hormone				
Prolactin				
Oxytocin				

UNIT THREE

How big is a trillion?

What is the role of a cell?

Why do I feel so bad in the morning
after the night before?

What is the difference between a virus
and a bacterium and how do I know
which one is attacking me?

CELL DIVISION . . . REALLY?

CELLS

EXAM III

Cell Membrane

Structures

Bacteria and Viruses

Cell Division

Cell Structure and Function

History

- **Anton van Leeuwenhoek**—17th century, "Father of the Microscope"
- **Robert Hooke**—coined the term "cell," from walled off chambers he found in cork samples
- **Mathais Schleiden**—1830s, German botanist that stated all plants were composed of cells

History

- **Theodor Schwann**—stated that all animals were also composed of cells
- **Rudolph Virchow**—added that all cells come from preexisting cells
- **Cell Theory**—combination of work of all four men

Cell Size

- **Surface Area/Volume Ratio**—nutrients enter and wastes exit through the membrane on the surface
- **Volume**—represents the needs of all cells
- **Surface Area**—represents the cell's ability to get material in and out of the cell

As the volume increases through growth, the surface area/volume ratio decreases

Surface Area/Volume Ratio

$$\frac{6\times1\times1}{1\times1\times} = \frac{6}{1} = \frac{sa}{v} = \frac{cm}{c} = \frac{sn}{n}$$

Cell Structure and Function
Cell Membrane

- Thin, nearly invisible structure that surrounds the cytoplasm
- Connects the endoplasmic reticulum with the nuclear membrane

Cell Membrane

Made up of two parts:

1. **Phospholipids**—with tails in the diagram
2. **Protein**—center of the diagram

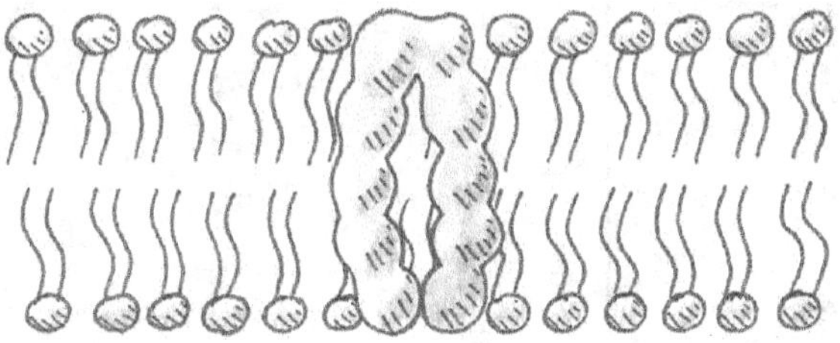

Cell Membrane

Three types of protein molecules found in the membrane:

1. **Transport**—regulate transport and diffusion
   - *Carrier Proteins*—move molecules one at a time through the lipid layer, they do not extend through the membrane

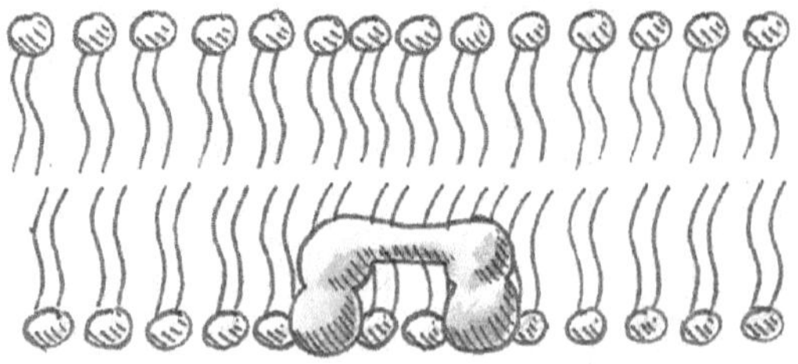

Image courtesy of Kevin Mercer

Cell Membrane

- *Channel Proteins*—extend through the lipid layer and allow for the random diffusion of molecules through the membrane

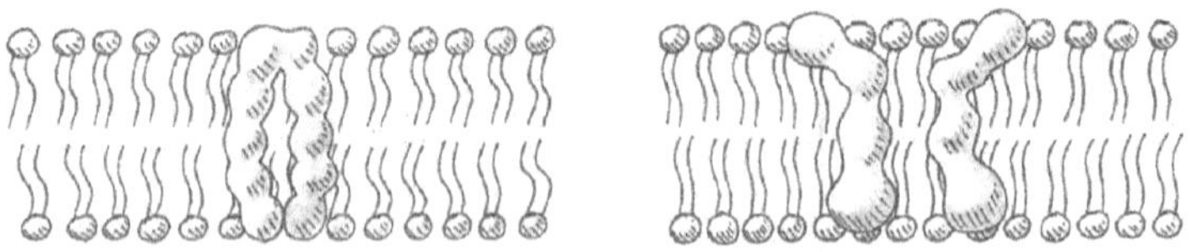

Image courtesy of Kevin Mercer

Cell Membrane

2. **Marker Proteins**—identify cells to other cells (immune system)

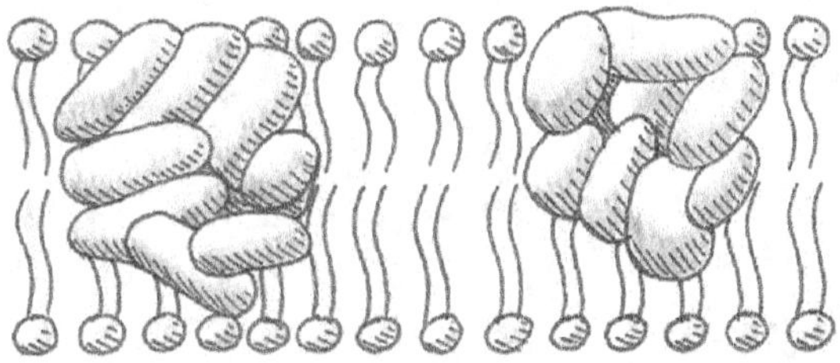

Image courtesy of Kevin Mercer

Cell Membrane

3. **Receptors**—allow the cell to receive instructions

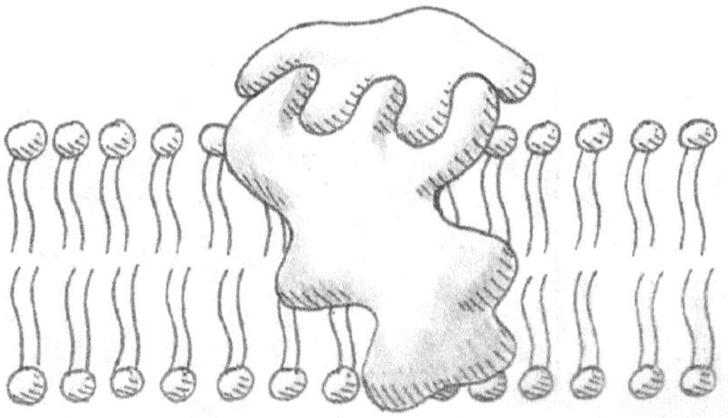

Image courtesy of Kevin Mercer

Cell Membrane Structure and Function

- Regulates the passage of molecules into and out of the cell
  - **Passive Transport**
    - *Diffusion*—needs a concentration gradient only, moves to lower concentration
  - **Active Transport**
    - *Facilitated Diffusion*—needs a carrier molecule and concentration gradient, moves toward lower concentration

Passive Transport Diffusion

- Movement of molecules from an area of higher concentration to an area of lower concentration (down the concentration gradient) until an equilibrium is achieved
- **Key Points:**
  - Always down the concentration gradient
  - Requires no chemical energy
  - Continues until equilibrium is reached

Osmosis
The Diffusion of Water into and out of Cells

- A differentially permeable membrane separates the cell and its environment
- A concentration gradient exists between the cell and its environment
- The differentially permeable membrane allows for the passage of a Solvent (water) but not Solute
- Process will continue until equilibrium is reached

Osmosis

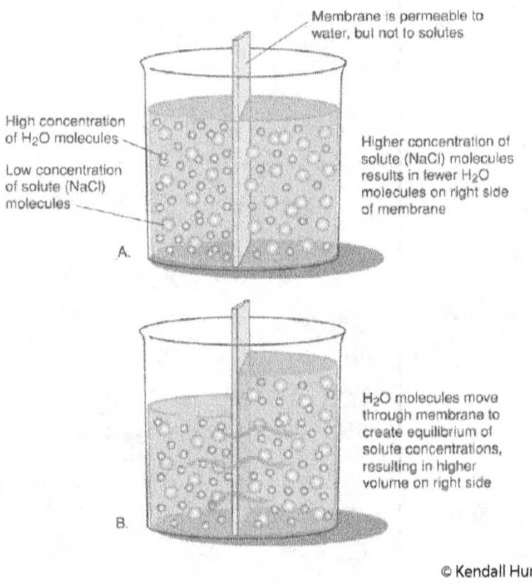

Osmotic Pressure

- The pressure that develops in a system due to osmosis
- The greater the osmotic pressure the greater the likelihood that water will diffuse in that direction

Tonicity

Isotonic Solution

- The solute concentration is the same on both sides of the membrane, therefore there is no net gain or loss of water
- Example:
  - A 0.9% solution of NaCl is known to be isotonic to red blood cells
  - Intravenous injections, medically administered

Tonicity

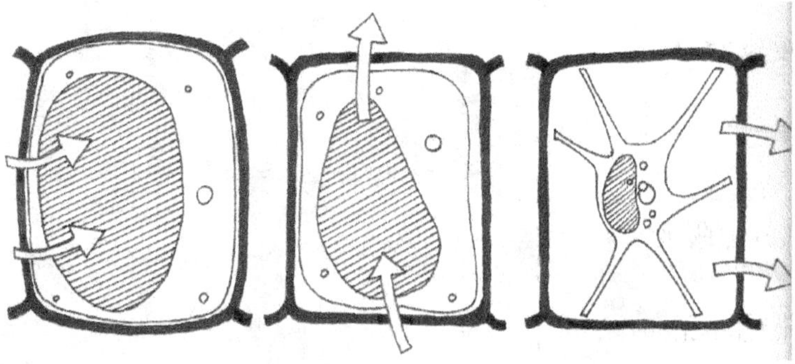

Image courtesy of Kevin Mercer

Hypotonic Solutions

- The prefix Hypo- means less than, and refers to a solution with a lower percentage of solute (more water) than the cell
  - A cell placed in a hypotonic solution will show a net increase in volume
  - There will be a net flow into the cell
- Example:
  - Any solution with a concentration lower than 0.9% is hypotonic to red blood cells
  - Animal blood cells placed in such a solution will swell and burst (Hemolysis)
  - Animal cells placed in such a solution will swell and burst (Plasmolysis)

Hypertonic Solution

- The prefix Hyper- means more than, and refers to a solution with more solute (less water) than the cell
  - A cell placed in a hypertonic solution will show a net decrease in volume
  - There will be a net flow of water out of the cell
- Example:
  - Any solution with a concentration higher than 0.9% is hypertonic to red blood cells
  - Animals placed in such a solution will shrink (Crenation)
  - Meats preserved by salting, the salt removes water from the meat killing the bacteria

Hypo-Hypertonicity

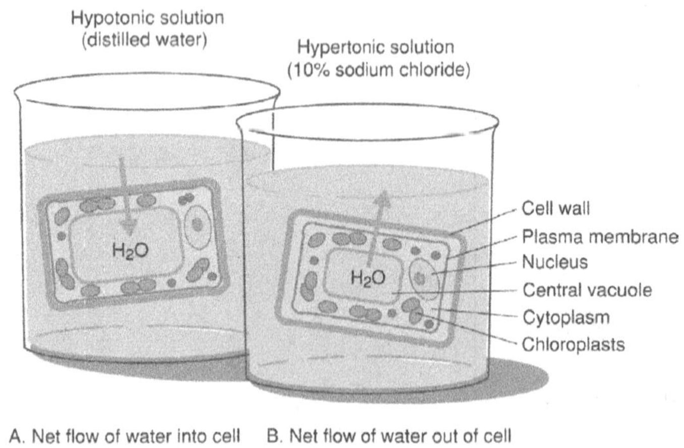

Transport by Carrier Proteins

Active Transport

- Carrier molecules are specific
- Explain the passage of amino acids and glucose across the plasma membrane
- Requires chemical energy
- Can occur as often as 100 times a second

Transport by Carrier Proteins

Active Transport

- Ions or molecules move through the plasma membrane, accumulating either inside or outside of the cell
- Therefore they are moving against their concentration gradient and expending cellular energy
- Estimates suggest that as much as 40% of a cell's energy is spent on active transport

Transport by Carrier Proteins
Active Transport

- **Requirements:**
  - Energy
  - Carrier Molecules
- **Example:**
  - Sodium-Potassium Pump:
    - Moves sodium ions (Na^+) to the outside of the cell
    - Moves potassium ions (K^+) to the inside of the cell

Exocytosis

- Requires cellular energy
- Vesicles formed by the Golgi apparatus
- Carry a specific protein
- Fuses with the plasma membrane as secretion occurs
- **Example:**
  - This is the way that insulin leaves insulin-secreting cells

Endocytosis

- Requires cellular energy
- Cell takes in substance by vesicle formation
- **Example:**
  - When the material taken in is large (food molecule) the process is called *Phagocytosis*
  - When the material taken in is a liquid or very small the process is called *Pinocytosis*

Cell Structure
Nucleus

- Regulates all cell activities
- Consists of two parts:
  - Nuclear envelope—outer membrane
  - Nucleoplasm—contains the chromatin (genetic material) and nucleolus (manufacture of ribosomes)

Nucleus

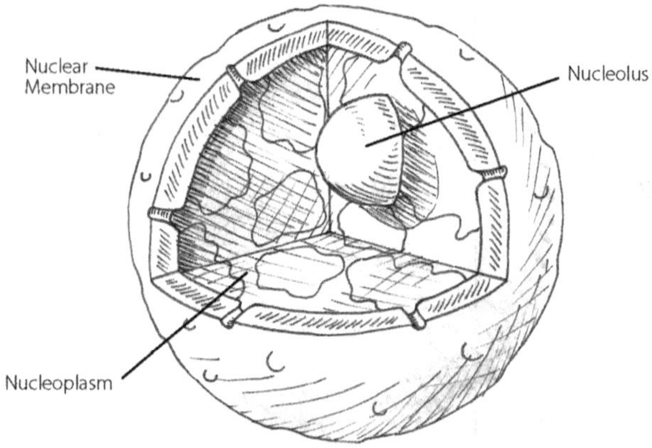

Cell Structure
Endoplasmic Reticulum

- Folded system of membranes that loop back and forth throughout the cytoplasm
- Provide a surface area for all cell reactions
  - Smooth—have no ribosomes associated with it (Lipid production)
  - Rough—has ribosomes associated with it (Protein production)

Cell Structure
Ribosomes

- Manufactures protein for the cell

Cell Structure
Golgi Body

- Responsible for packaging proteins for the cell
- Produces vacuoles and vesicles

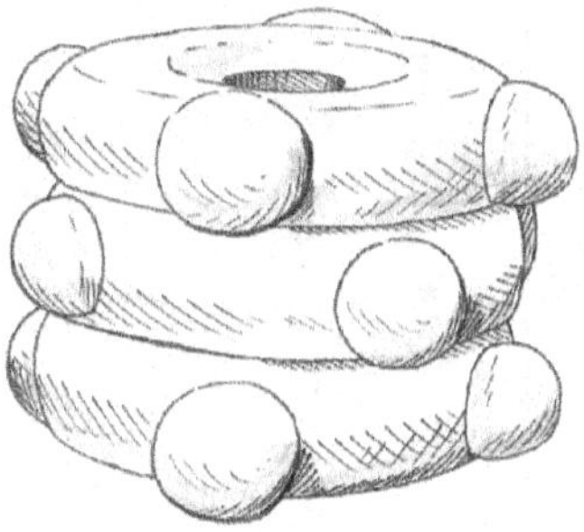

Cell Structure

Lysosomes

- Produced by the Golgi body
- Special kind of vacuole
- Consists of a single membrane surrounding powerful digestive enzymes

Mitochondrion

- The site of aerobic respiration
- Glucose + oxygen = carbon dioxide + water + ATP

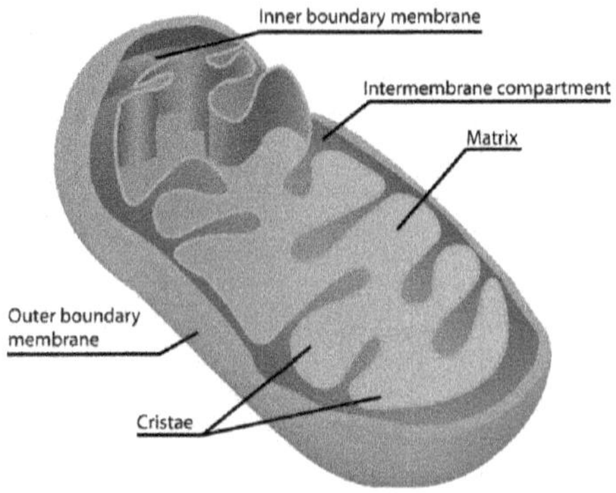

Cell Structure

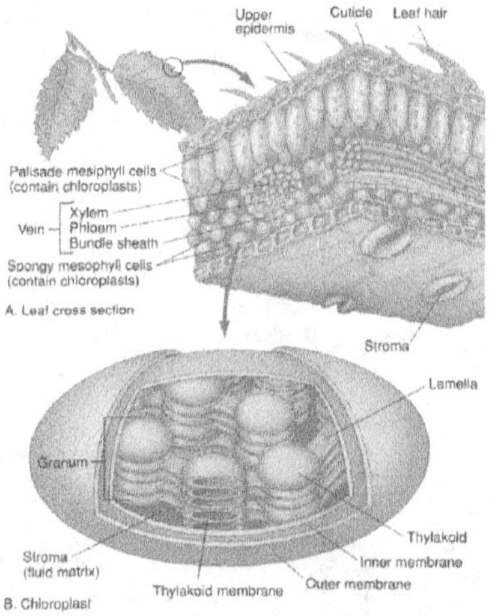

Chloroplast

- Found only in plant cells
- Site of photosynthesis (Light energy is converted into food for the plant)

Cell Structure

Centrioles

- Found only in animal cell
- Functional in cell division
- Oriented at right angle to each other

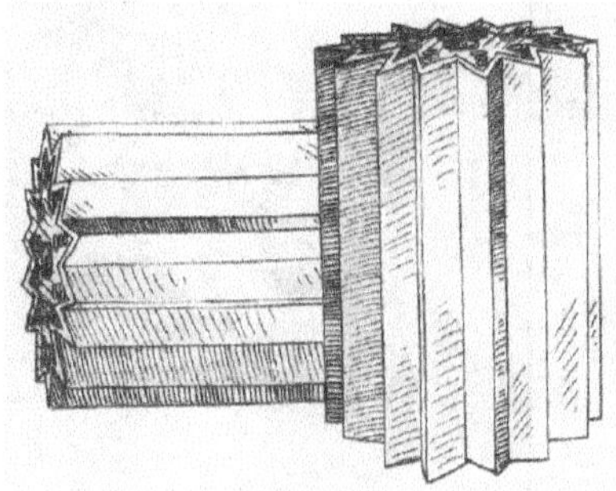

Cell Structure
Vacuoles

- Large empty appearing areas found in the cytoplasm
- Usually found in plant cells where they store waste, food, water, etc.

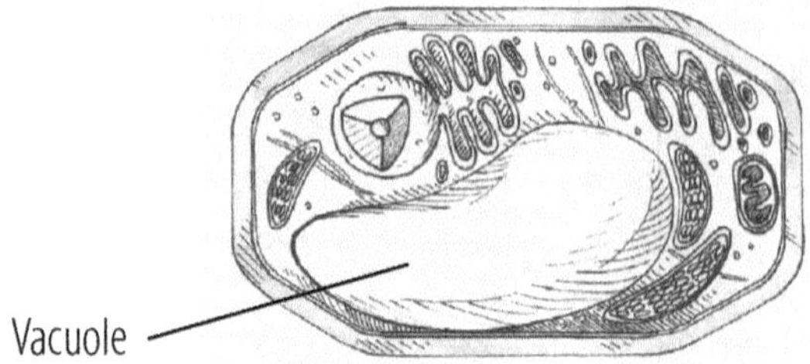

Cell Structure
Cell Wall

- Found only in plant cells and bacteria
- Contains cellulose fibrils
- Provides support and protection

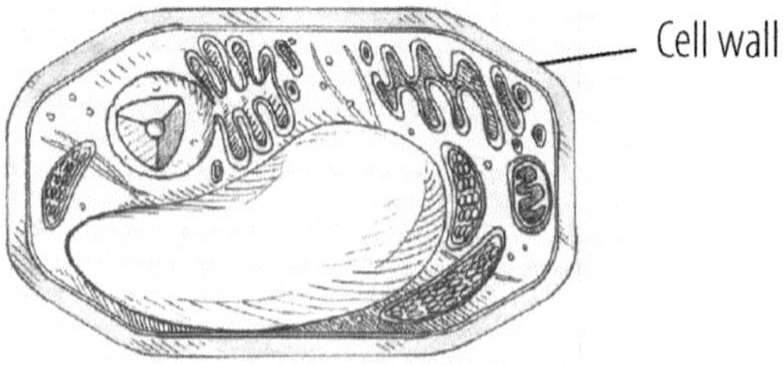

Cell Structure
Cytoskeleton

- A network of interconnected filaments and tubules that extends from the nucleus to the plasma membrane
- Maintains cell shape
- Causes the cell and its organelles to move

Cell Structure
Cytoskeleton: Actin Filaments

- Long, thin fibers
- Occur in bundles or mesh-like networks
- Provide a structural function when they appear as a network under the plasma membrane
- Play a role in cellular movement when it bundles

Cell Structure

Cytoskeleton: Microtubules

- Small, hollow cylinders
- Help maintain the shape of the cell and provide tracks along which organelles can move

Cell Structure

Cilia

- Short, stiff, hair-like projections
- Appear as beating oars, working in unison
- Mode of locomotion in single celled organisms (ciliates)
- Found in our respiratory tract, sweep debris from our lungs back up to our throat to be swallowed

Cell Structure
Flagella

- Long, flexible, hair-like projections
- Appear as an undulating whip
- Allow for motility in mammalian sperm and single-celled organisms (Flagellates)

Cell Structure
Flagella

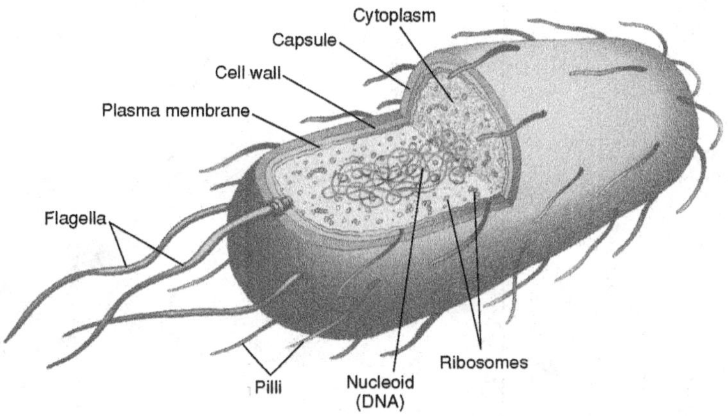

Taxonomy

Binomial Nomenclature

- Carolus Linnaeus (1707–1778)
- Swedish biologist
- Invented this two-part naming system

Binomial System

Made up of two names:

1. Genus
2. Species

Binomial Rules

1. **Genus**
   - Always first
   - Always capitalized
   - Must be *italicized* or underlined

Binomial Rules

2. **Species**
   - Always second
   - Always lower case
   - Must be *italicized* or underlined
   - **Example:**
     - *Homo sapiens* or Homo sapiens

Archae (Archaebacteria)

- Cell walls lack peptidoglycan
- Characteristic of all bacteria
- Contain very unusual lipids
- Characteristic Ribosomal RNA (rRNA) sequence
- Some genes have Introns
  - Unlike those of bacteria

Extremophiles

- **Thermophiles (Heat Lovers)**
  - Live in very hot places, like hot springs
  - Typically from 60°–80° C
    - As high as 113° C
  - Many have metabolism based on sulfur
- **Pressure Tolerant**
  - Some need 300 atms to survive and can thrive up to 800 atms

Extremophiles

- Halophiles (Salt Lovers)
  - Live in water with salinity of 15% to 20%
  - Seawater is 3%
- pH Tolerant
  - Live in areas with very high pH (Basic) or very low pH (Acidic)

Bacteria (Eubacteria)

- Prokaryotic
- Single celled
- No organized nucleus
- Must numerous of all domains

Bacterium Anatomy

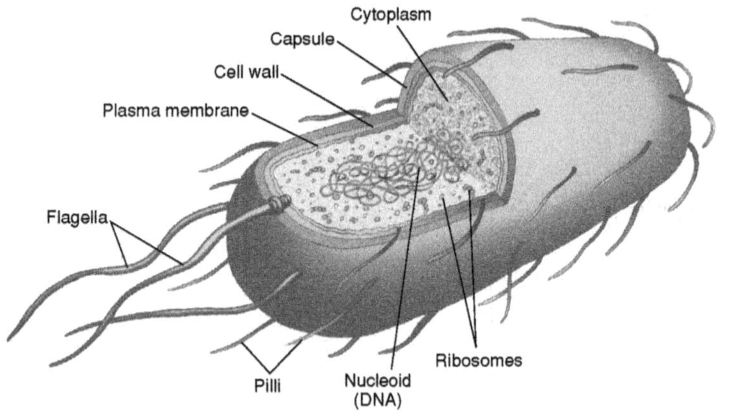

Prokaryotes

- Compose two kingdoms or domains (Archaea and Eubacteria)
- Are cellular
- Typically 1/1000 the volume of small eukaryotic cells
- Most are unicells, but some are colonies or filaments

Bacteria

Shapes

- **Spherical (cocci)**
  - Diplococci—groups of two cocci
  - Streptococci—chains of cocci
  - Staphylococci—groups of cocci
- **Rod shaped (bacilli)**
  - Can occur singularly or in chains
- **Vibrios**—short, spiral-shaped
- **Spirilla**—long, spiral-shaped and rigid
- **Spirochetes**—long, spiral-shaped and flexible

Bacteria

- Have ribosomes
- Lack all membrane-bound organelle
- Plasma membrane may be infolded to increase the surface area for enzymes needed for respiration and photosynthesis
- Cell surface is generally covered by a cell wall

autism spectrum disorder (asd)

· usually shows up between 2-3 years of age

· causes - genetic and environmental

Symtoms - Can't make eye contact, language development, repeative behavior

Bacteria

- Cell wall
- Very strong
- Resists bursting (adapted to living in a hypotonic media) do not grow well in hypertonic media
  - Hence, many of our methods of preserving food
  - High in sugar or salt
  - Act to prevent bacterial growth

Bacteria
Christian Gram

- Developed the "Gram Stain" in 1888
- Bacteria that stain purple (gram positive) due to thick peptoglycan walls
- Bacteria that remain red in color and do not absorb the purple stain (gram negative) have a thick outer layer of lipids and carbohydrates
- Penicillin interferes with peptidoglycan synthesis and is therefore most effective against gram-positive bacteria

Bacteria
General Morphology

- **Capsule**—protects against phagocytosis or attachment to substrate
- **Pili**—used for attachment or transmission of DNA between bacteria
- Many types are motile

Bacteria
Single DNA Molecule

- Located in the cytoplasm (not surrounded by a membrane)
- It is circular and highly folded
- **Plasmids**—extraneous DNA circles which replicate independently of the primary DNA
  - Typically have genes that allow resistance to antibiotics

Bacteria
Asexual Reproduction

- Binary Fission:
  - Asexual process
  - Circular DNA replicates
  - A new cell wall grows between the DNA molecules
- Can occur every 20 minutes under optimal conditions

Bacteria
Pseudo-Sexual Reproduction

- Not sexual reproduction
- Creates an individual with a new genetic code
- Genetic material may transfer through three mechanisms
- **Transformation**—fragments of DNA from a damaged cell may be taken up by another cell

Bacteria
Pseudo-Sexual Reproduction

- **Transduction**—bacteriophages (virus) transfer bacterial DNA from one to the other
- **Conjugation**—genetic material may be transferred via pili
  - DNA is transferred between different mating strains
  - Exchange occurs between donor and recipient strains

Viruses

- An infectious agent not assigned to any of the six kingdoms
- Consists of a nucleic acid surrounded by a protein coat (Capsid)
- Some also have an outer membranous envelope
- Contain either DNA or RNA, not both

Typical Virus

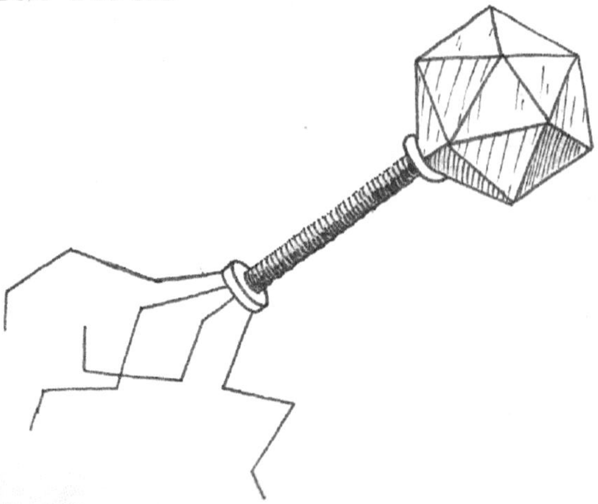

Viruses

Bacteriophages

- Viruses that attack bacteria
- Complex viruses that consist of nucleic acid (typically DNA)
- Enclosed in a polyhedral head
- Usually with a tail for attachment to the host
- Phages are important in genetic research

Viruses
Lytic Replication Cycle

- Destroys the cell
- **Attachment** (or absorption) is the process by which the phage attaches to the bacterial cell wall
- **Penetration**—is accomplished by the tail of the virus pushing a hole in the cell wall of the host and injecting its nuclear material into the bacterial cell

Viruses
Lytic Replication Cycle

- The capsid remains outside of the bacterium
- **Replication**—the phage utilizes the host cells ribosomes, energy, and enzymes to synthesize raw materials
- The synthesized macromolecules are assembled into new phages
- The host cell *Lyses* and releases new phages
- The process takes about 30 minutes

Replication Cycle

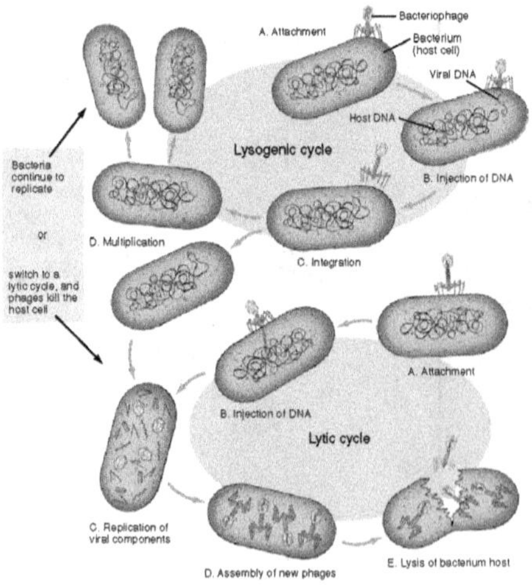

Lysogenic Replication Cycle
Temperate Viruses

- Do not always lyse their host cells
- Viral DNA becomes incorporated into the DNA of the host
  - It is then called a *Prophage*
- The prophage becomes replicated when the host cell replicates

Lysogenic Replication Cycle

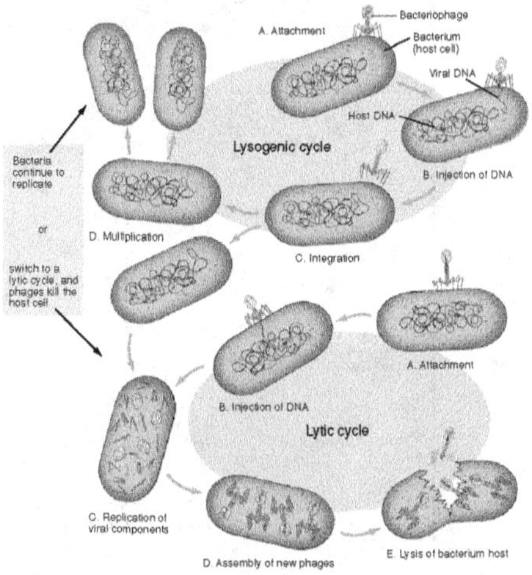

Lysogenic Replication Cycle
Temperate Viruses

- Lysogenic conversion occurs when the host cell then exhibits new or unusual properties
- The bacterium that causes diphtheria only causes the disease when it contains a certain phage
- The bacterium that causes botulism is harmless unless it carries a particular prophage

Viruses
Animal Cell Infections

- Hundreds of viruses infect animal cells
- Receptor sites are typically species specific and often tissue specific
- May enter by fusion with the plasma membrane or endocytosis
- May contain either DNA or RNA

Viruses
Animal Cell Infections

- DNA viruses synthesize new nucleic acids and proteins in a manner similar to typical synthesis
- Some RNA viruses are *Retroviruses*
- HIV and some cancer-causing viruses are of this type
  - that use an enzyme called *Reverse Transcriptase*
  - to transcribe RNA into a DNA intermediate
  - which is then used to make new viral RNA

Viruses
Animal Cell Infections

- A typical human suffers from 2–6 viral infections per year
- **Example:**
  - Human Viral Diseases—
    - Common cold
    - AIDS
    - Ebola
    - Herpes
    - Measles
    - Hepatitis
    - Mumps
  - Other Animals—
    - Canine distemper
    - Feline leukemia
    - Foot-and-mouth

Viruses
Plant Cell Infections

- May be spread to plants by fluid-feeding insects (aphids) or by infected seeds or asexual reproduction
- Spread through the plant via plasmodesmota
- Most are RNA based
- Infected plants show stunted growth, spots on the leaves, and lower yields
- Treatments are not well developed

Viroids and Prions

- **Viroids**—Contain no proteins or genes to code for proteins
- **Prions** (proteinaceous infectious particles)—are composed of protein but contain no nucleic acids
- Responsible for
  - Transmissible spongiform encephalopathies (Scrapie in sheep and goats)
  - Bovine spongiform encephalopathy "Mad cow disease"
  - Variant Creutzfeldt-Jakob disease in humans
  - Chronic Wasting Disease in the Elk and Deer herds

Eukarya (Eukaryotes)

- Appear about 1.5 BYA
- Include the following four Kingdoms
  - Fungi
  - Plants
  - Animals
  - Protists

Eukaryotes

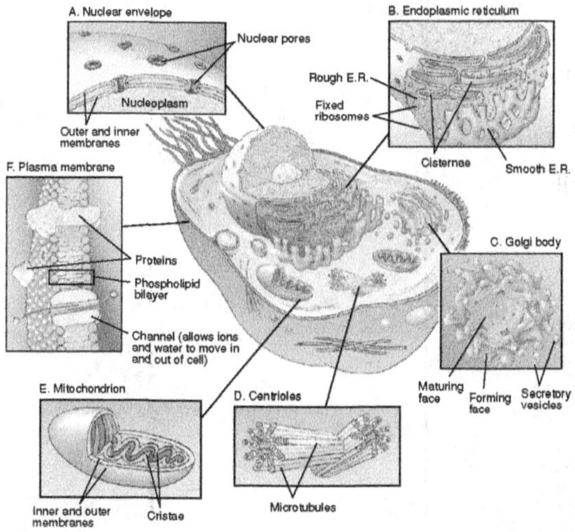

Classification of Man

Domain	Eukarya	Denmark's
Kingdom	Animalia	King
Phylum	Chordata	Phil
Subphylum	Vertebrata	Sometimes
Class	Mammalia	Comes
Order	Primates	Over
Family	Hominidae	For
Genus	*Homo*	Gene's
Species	*sapiens*	Special

Cell Division
Maintaining the Chromosomal Number

- **Chromatin**—tangled mass of thin threads of DNA found in the nucleus (when the eukaryotic cell is not undergoing cell division)
  - Looks like angel hair pasta
- **Chromosomes**—chromatin coils, loops, and condenses to become these highly compacted structures (during cell division)
  - Looks like macaroni

Cell Division

- **Cell Division**—involves nuclear division and cytokinesis
- **Cytokinesis**—division of the cytoplasm
- **Mitosis** nuclear division in which the chromosome number remains constant
  - Occurs in growth and repair of the body
- **Example:**
  - Diploid = Diploid and Diploid
  - Haploid = Haploid and Haploid

The Cell Cycle
Interphase

- Made up of all the stages of the cell cycle except mitosis and cytokinesis
- G1—Organelles begin to double in number
- S—Replication of DNA
- G2—Synthesis of proteins
- M—Mitosis and cytokinesis

Cell Cycle

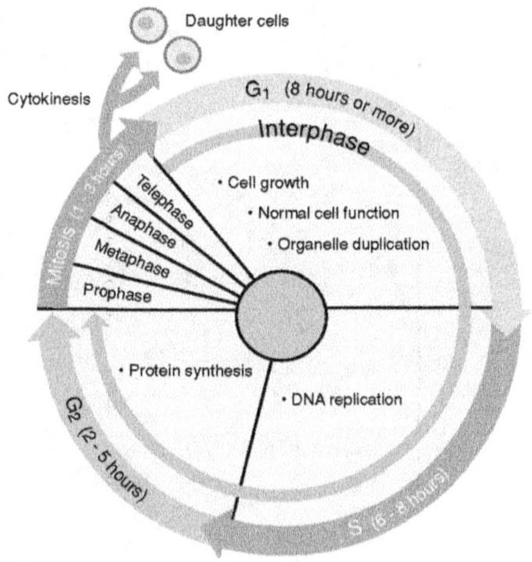

Interphase–Mitosis

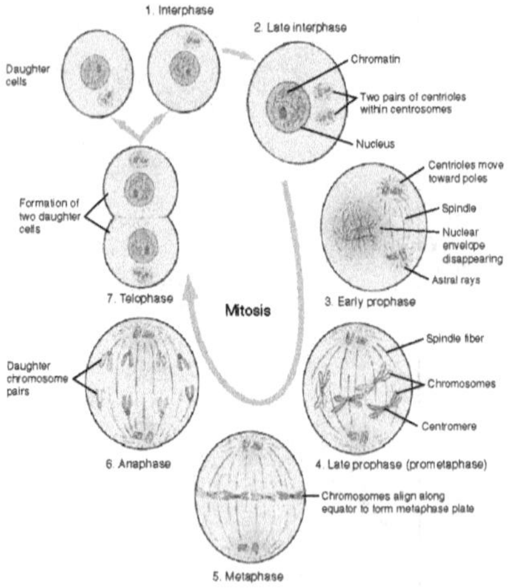

Mitosis
Prophase

- Centrosomes begin moving to opposite ends of the nucleus
- Spindle fiber appear between the separating centrosomes
- Nuclear envelope starts to fragment
- Nucleolus starts to disappear
- Spindle fibers attach to the chromosomes

Mitosis
Metaphase

- Spindle is now fully formed
- Chromosomes align at the Metaphase Plate (Equator)

Mitosis
Anaphase

- Centromeres divide
- Sister chromatids separate to become two daughter chromosomes
- Chromosomes move to opposite end of the cell

Mitosis

Telophase

- Spindle disappears
- Nuclear envelope components reassemble around the daughter chromosomes
- Nucleolus reappears
- Ends with the separation of the two daughter cells

Plant Cells

- **Cell Plate**—forms in the middle of the cell and grows toward the edges, dividing the two daughter cells

Animal Cells

- **Cleavage Furrow**—appears (indentation of the cell membrane) and continues to deepen between daughter cells until division is complete

Meiosis

Formation of Gametes

"Reduction Division"

- Occurs in any life cycle that involves sexual reproduction
- Reduces the chromosome number
- Daughter nuclei receives only one member of each homologous pair
- Process ensures that the next generation will have a combination of traits that are different than the parents

Meiosis I
Prophase I

- The spindle appears while the nuclear envelope fragments
- Nucleolus disappears
- Homologous chromosomes each have two sister chromatids
- Synapsis takes place, allowing crossing-over (the sister chromatids are no longer identical)

Meiosis I
Metaphase I

- Homologous pairs are aligned at the equator

Meiosis I

Anaphase I

- Homologous chromosomes separate and move to opposite poles of the spindle
- Each chromosome consists of two sister chromatids

Meiosis I

Telophase I

- Daughter cells separate
- No replication of DNA occurs during this period of time between divisions called *Interkinesis*

Meiosis

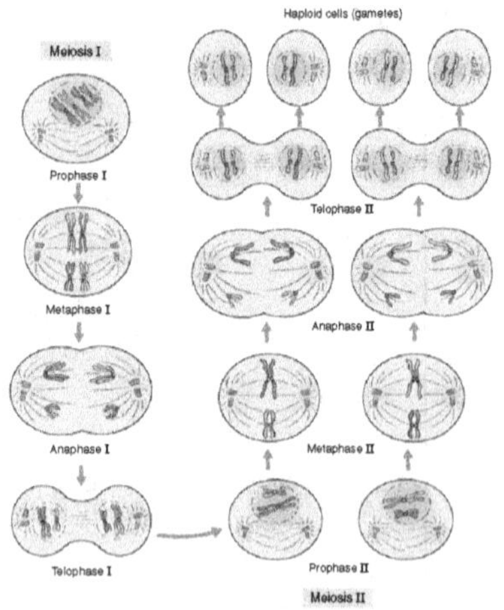

Meiosis II
Prophase II

- Spindle appears while the nuclear envelope disassembles
- Nucleolus disappears
- Each duplicate chromosome is attached to the spindle

Meiosis II
Metaphase II

- Each chromosome lines up independently at the equator

Meiosis II
Anaphase II

- Centromeres divide
- Sister chromatids separate becoming "Daughter Chromosomes"

Meiosis II

Telophase II

- Spindle disappears
- Nuclear envelopes re-form
- Each daughter cell is now **haploid**
- Four daughter cells are produced

Contrasts

Meiosis I and Mitosis

- There is no crossing-over in mitosis
- Paired homologous chromosomes align at the equator in meiosis, while individual duplicate chromosomes align at the equator in Mitosis

Contrasts

Meiosis II and Mitosis

- DNA replication only takes place once in both; however, there are two nuclear divisions during meiosis and only one in mitosis
- Four "Daughter Cells" are produced by meiosis while only two are produced by mitosis
- Meiosis produces haploid daughter cells and diploid daughter cells are produced by mitosis

Mitosis vs. Meiosis

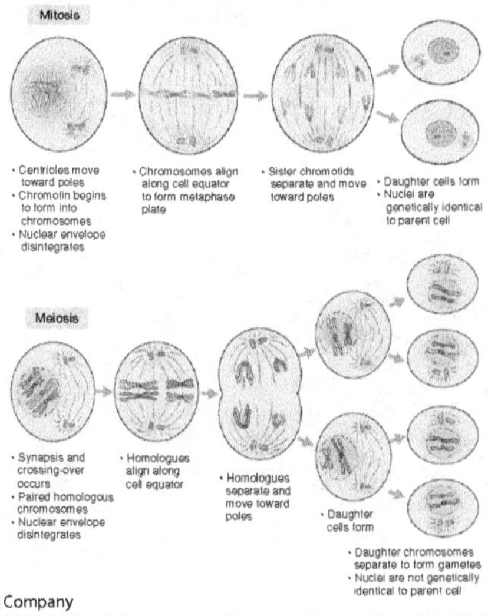

UNIT FOUR

Who am I?

Why do I look more like my great, great grandfather . . . "Daddy Fisher" than mom or dad?

Where did I come from?

How do I ensure passing on only my "good genes" to the next generation?

Why can't anybody teach me to roll my tongue?

GENETICS

EXAM IV

Introduction and History

Problems

Disorders

Molecular

Foundation of Genetics

- **Gregor Mendel**—Austrian monk, developed laws of inheritance in 1860
- **Gene**—unit of heredity that codes for a protein and is passed on to the offspring
- **Alleles**—alternative forms of a gene having the same position on a pair of homologous chromosomes and affecting the same trait
- **Loci**—location of the gene on the chromosome

Mendel's Laws
The Law of Segregation

- Each individual has two factors for each trait
- The factors segregate (separate) during the formation of the gametes
- Each gamete contains only one factor from each pair of factors
- Fertilization gives each new individual two factors for each trait

Independent Assortment

- Each pair of factors segregates (assorts) independently of the other pair
- All possible combinations of factors can occur in the gametes
    - Dependent on the random arrangement of the homologous pairs at the metaphase plate of the spindle

Mendelian Cross

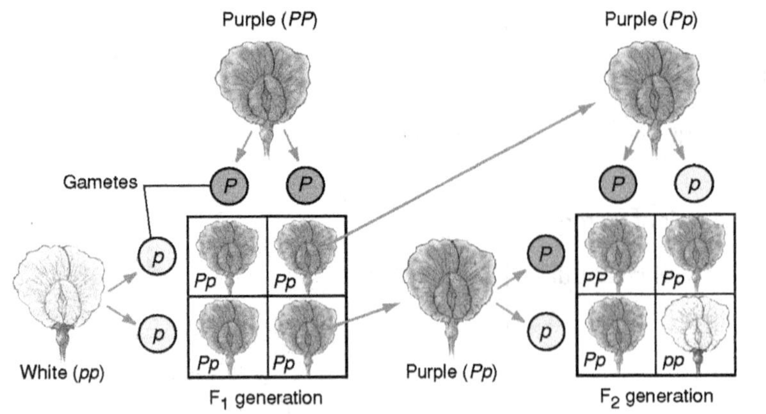

Gamete Formation

- The gametes have only one allele for each trait in accordance with Mendel's Law of Segregation
- The member of each allelic pair separate during meiosis
- Gamete (haploid) therefore contain exactly half the number of alleles as the parents cells (diploid)

Gamete Formation
Practice Problems

- How to determine the possible gametes that could be produced from an individual of known genotype . . .
- **Example:**
  RrWWTtbb = Parental genotype
  (RWTb) (rWTb) (RWtb) (rWtb) = Possible gametes
- **Example:**
  RRWwttBB = Parental genotype
  (RWtB) (RwtB) = Possible gametes

Single Trait Inheritance

- Mendel suggests using letter to represent factors now called alleles
- **Dominant Trait**—indicated by the use of a representative capital letter
- **Recessive Trait**—indicated by the use of the appropriate lower case letter

Genotype

- Two letters used to represent the genes of the individual
- **Homozygous Recessive**—the two members of the allelic pair are the same and both recessive (ww)
- **Homozygous Dominant**—the two members of the allelic pair are the same and both dominant (WW)
- **Heterozygous**—the two members of the allelic pair are different (Ww)

Phenotype

- The physical characteristics of the individual—what the individual looks like as well as microscopic and metabolic characteristics
- **Example:**

Phenotype:	**Genotype:**
Widow's peak	WW
Widow's peak	Ww
Straight hairline	ww

One-Trait Crosses
Monohybrid

- Parental generation (P_1)—Ww × Ww
- Gametes (W) (w) (W) (w)
- Punnett square

	W	w
W	WW	Ww
w	Ww	ww

One-Trait Crosses
Monohybrid

- First filial generation (F_1)
- Genotype: ¼ WW, ½ Ww, ¼ ww
- Phenotype: ¾ widow's peak, ¼ straight hairline
- **Dominant Allele**—masks the expression of the recessive allele
- **Recessive Allele**—is expressed only when no dominant allele is present

Monohybrid Cross

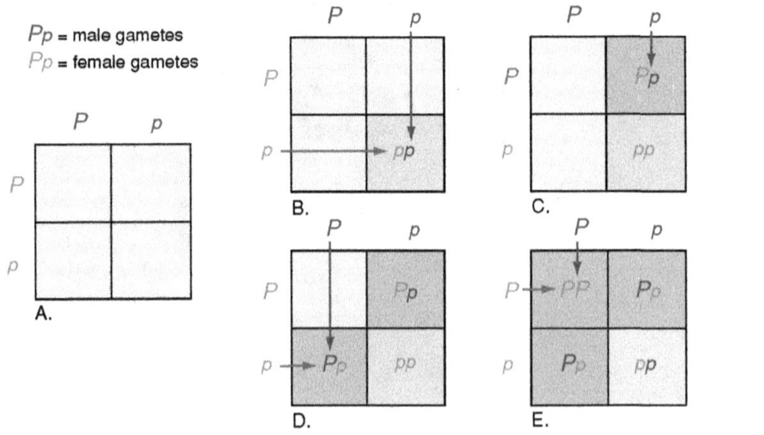

Probability

- $P_1 = Ww \times Ww$
- Gametes:
  The chance of (W) = ½
  The chance of (w) = ½
- Genotype:
  WW = ½ × ½ = ¼
  Ww = ½ × ½ = ¼
  wW = ½ × ½ = ¼
  ww = ½ × ½ = ¼

One-Trait Crosses
Monohybrid Cross

- **Test Cross**—occurs when an individual with the dominant phenotype is crossed with an individual having the recessive phenotype
  - Used to determine the genotype of the parent with the dominant phenotype

Test Cross

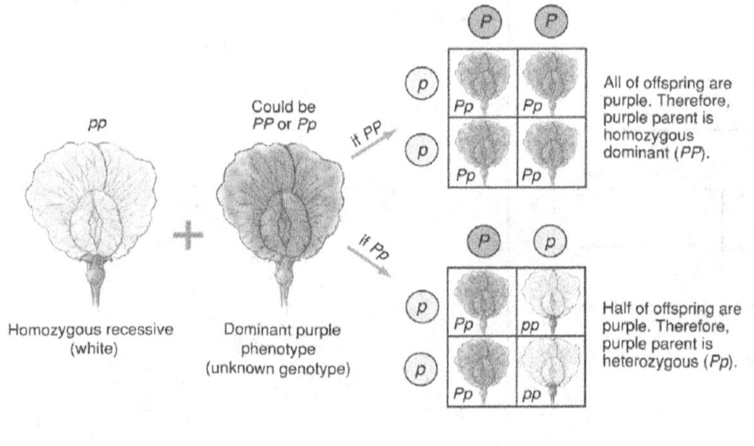

Two-Trait Crosses

W = Widow's peak   w = Straight hairline

S = Straight thumb   s = Hitchhiker's thumb

P_1 = WWSS × wwss

Gametes = (WS) (ws)

F_1 = WwSs

Gametes = (WS) (Ws) (wS) (ws)

Dihybrid Cross
Punnett Square

	WS	Ws	wS	ws
WS	WWSS	WWSs	WwSS	WwSs
Ws	WWSs	WWss	WwSs	Wwss
wS	WwSS	WwSs	wwSS	wwSs
ws	WwSs	Wwss	wwSs	wwss

Dihybrid Cross

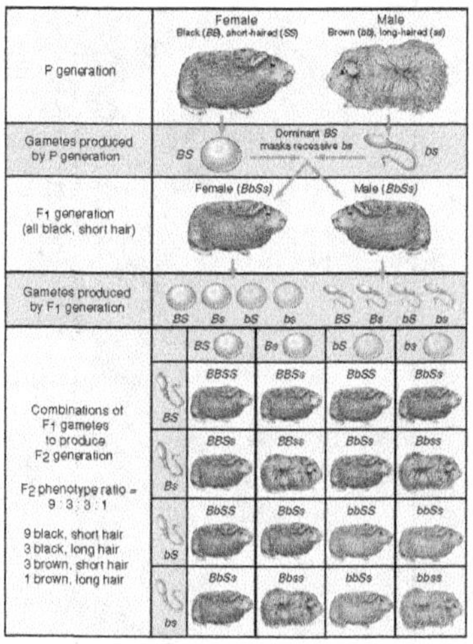

Two-Trait Cross and Probability

- Phenotype:

  ¾ = Widow's peak ¼ = Straight hairline

	W	w
W	WW	Ww
w	Ww	ww

Two-Trait Cross and Probability

- Phenotype:

  ¾ = Straight thumb ¼ = Hitchhiker's thumb

	S	s
S	SS	Ss
s	Ss	ss

Two-Trait Cross and Probability

Therefore:

- The probability of the offspring having:
  - Widow's peak and straight thumb = ¾ × ¾ = 9⁄16
  - Widow's peak and hitchhiker's thumb = ¾ × ¼ = 3⁄16
  - Straight hairline and straight thumb = ¼ × ¾ = 3⁄16
  - Straight hairline and hitchhiker's thumb = ¼ × ¼ = 1⁄16
- Phenotypic Ratio = 9:3:3:1

Genetic Disorders
Patterns of Inheritance
Pedigree

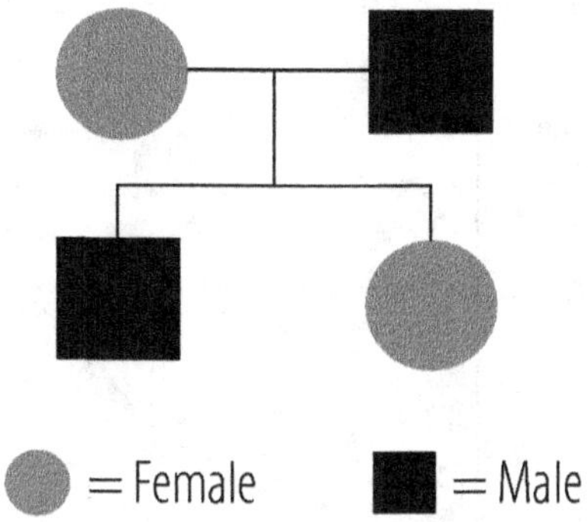

Genetic Disorders
Patterns of Inheritance
- Shaded circles or squares are affected or posses the trait in question
- **Carriers**—if the trait in question is recessive, individuals appear normal (heterozygotic) but are capable of producing offspring that have the disorder

Albinism-Pedigree

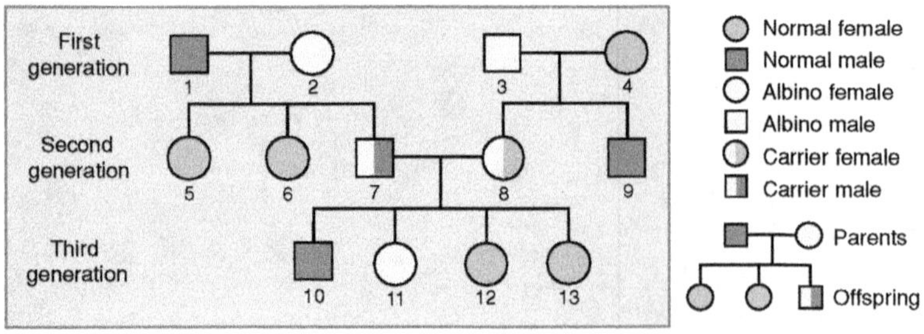

Genetic Disorders

Autosomal Dominant Disorders

- Affected children usually have an affected parent
- Heterozygotes (Aa) are affected as well as homozygotes (AA)
- Two affected parents can produce an unaffected child
- Two unaffected parents will not have affected children
- Both males and females are affected with equal frequencies

Autosomal Dominant Disorder

Neurofibromatosis (von Recklinghausen Disease)

- One of the most common genetic disorders
- Six or more large tan spots appear on the skin
- May increase in size and number or become darker with age
- Small benign tumors may occur under the skin
- In most cases symptoms are mild and patients have a normal life

Autosomal Dominant Disorder

Huntington Disease

- A neurological disorder that leads to progressive degeneration of brain cells
- Causes severe muscle spasms and personality disorders
- Most people appear normal until middle age and have already had children
- There are no treatments
- Death occurs 10–15 years after onset of symptoms

Autosomal Recessive Disorder

- Most affected children have normal parents
- Two affected parents will always have an affected child
- Affected individuals with homozygous normal mates will have normal children
- Close relatives who reproduce are more likely to have affected children
- Both males and females are affected with equal frequency

Autosomal Recessive Disorder
Tay-Sachs Disease

- Usually occurs among Jewish people in the U.S.
- Development slows between 4 and 8 months of age
- Neurological impairment and psychomotor difficulties become apparent
- The child gradually becomes blind, develops uncontrollable seizures eventually becoming paralyzed
- There is no cure

Autosomal Recessive Disorder
Cystic Fibrosis

- Most common lethal genetic disease among Caucasians in the U.S.
- Mucus in the bronchial tubes and pancreatic ducts is particularly thick and viscous
- Interfering with the function of the lungs and pancreas
- Life expectancy is 17–28 years

Autosomal Recessive Disorder
Phenylketonuria (PKU)

- Lacks an enzyme that is needed for the normal metabolism of the amino acid Phenylalamine
- An abnormal breakdown product (Phenylketone) accumulates in the urine
- Newborns are placed on a diet low in Phenylalamine
- Special diet continues until brain is fully developed or severe mental retardation results

Polygenetic Inheritance
Epistasis

- Occurs when one trait is governed by two or more sets of alleles, possibly located on many different chromosomes
- Each dominant allele has a quantitative effect on the phenotype and these effects are additive
- The result is a continuous variation in phenotypes

Polygenetic Inheritance
Skin Color

- Unknown number of pairs of alleles (possibly as many as 13)
- Few people have extreme phenotype
- Most people have the phenotypes that lie in the middle

Polygenetic Disorders

- Cleft lip/palate
- Club foot
- Congenital dislocation of the hip
- Hypertension
- Diabetes
- Schizophrenia
- Allergies
- Cancers
- These may be subject to environmental influences

Multiple Alleles

- A trait is controlled by three alleles
- Each person only has two of the three possible alleles

Multiple Alleles
ABO Blood Type

- A = A antigen on the red blood cells
- B = B antigen on the red blood cells
- AB = A and B antigen on the red blood cells
- O = neither A nor B antigen on the red blood cells

Phenotype:	Possible Genotype:
A	I^AI^A, I^Ai
B	I^BI^B, I^Bi
AB	I^AI^B
O	ii

Rh Factor
Simple Mendelian Inheritance

- Inherited separately from A, B, AB, or O blood types
- Rh positive (Rh^+) is dominant over Rh negative (Rh^-)
- Rh^+ has a particular antigen on the red blood cells
- When an Rh^- woman reproduces with an Rh^+ man and the child in the womb is Rh^+ the woman may begin to produce antibodies that will attack the red blood cells of the baby
- This can be prevented by giving the mother Rh immunoglobin injections midway through the pregnancy

Degree of Dominance

- Codominance—ABO blood types
- Incomplete Dominance:
  - HH = Curly hair
  - HH' = Wavy hair
  - H'H' = Straight hair

Incomplete Dominance

Sickle-Cell Disease

Genotype:	Phenotype:
$Hb^A\ Hb^A$	Normal
$Hb^A\ Hb^S$	Sickle-cell trait
$Hb^S\ Hb^S$	Sickle-cell positive

Incomplete Dominance
Sickle-Cell Disease

- Sickle-shaped blood cells cannot pass along the narrow capillary passageways, clogging the vessels
- 60% of the population in malaria-infected regions of Africa have the allele
- 10% of the African-American population carry the allele

Incomplete Dominance
Sickle-Cell Disease

- Among regions of malaria-infected Africa—
    - infants with sickle-cell trait have a better chance of survival than the normal homozygote
    - the protection afforded by the sickle-cell trait keeps the allele prevalent in populations exposed to malaria

Patterns of Chromosomal Inheritance

- **Karyotype**—display of chromosome pairs derived from a photograph taken of a cell before division
- **Sex Chromosome**—contains the gene that determines the sex of the offspring (one pair)
- **Autosomes**—include all of the chromosomes except the sex chromosomes (22 pairs)

Nondisjunction

- Occurs during Meiosis I when a members of a homologous pair both go into one of the daughter cells
- Occurs in Meiosis II when the sister chromatids fail to separate and both daughter chromosomes go into the same gamete
- The zygote will have one more chromosome than usual (**trisomy**) or one less than usual (**monosomy**)

<u>**All autosomal monosomies and almost all trisomies are lethal**</u>

Downs Syndrome

Trisomy-21

- The most common autosomal trisomy
- Chances increase greatly in mothers over 40
- **Characteristics:**
  - Short stature, eyelid fold, stubby finger, wide gap between the first and second toes, large fissured tongue, round head, palm crease, and mental retardation

Chromosomal Mutations

A change in chromosome structure that can be detected microscopically

- **Inversion**—a segment of the chromosome is turned around 180°
- **Translocation**—the movement of a chromosome segment from one chromosome to another, nonhomologous, chromosome
- **Deletion**—an end of a chromosome breaks off or two simultaneous breaks lead to the loss of a segment
- **Duplication**—the presence of a chromosome segment more than once in the same chromosome

Sex Chromosomes

- **Barr Bodies**—a darkly staining mass of condensed chromatin found in the nuclei (inactive X chromosome)
- Normal females have one Barr body, normal males have none
- Counting the number of Barr bodies can help detect an abnormal number of X chromosomes

Abnormal Sex Chromosome Inheritance

Abnormal Number Syndromes

- Turner's Syndrome
  - Has only one sex chromosome
  - Genotype = XO
  - **Characteristic phenotype:**
    - All females, sexual organs never develop and are nonfunctional, short with a broad chest
    - Infertile even with hormone implants

Abnormal Sex Chromosome Inheritance

Abnormal Number Syndromes

- Klinfelter Syndrome
  - Has two or more X chromosomes in addition to a Y chromosome
  - Always male (Genotype XXY or XXXY)

Klinfelter Syndrome

Characteristics

- Testes and prostate gland are underdeveloped
- There is no facial hair
- May be some breast development
- Large hands and feet
- Very long arms and legs
- Slow learners, but, not usually mentally retarded unless they inherit more than two X chromosomes

Triplo-X Syndrome
Characteristics

- Always female
- Contain more than two X chromosomes and extra Barr bodies in the nucleus
- In some cases there is a tendency toward learning disabilities
- Most have no apparent physical abnormalities, except menstrual irregularities and early onset of menopause

Jacob's Syndrome
Characteristics

- Genotype XYY
- Usually taller than average
- Persistent acne
- Speech and reading problems

Rumors of increased criminal activity were not supported by scientific research

Sex-Linked Inheritance

Genes found on the sex chromosomes that are not involved with sexual development

- **X-linked:** found on the X chromosome
- **Y-linked:** found on the Y chromosome
- Most are X-linked because the X chromosome is so much larger than the Y chromosome

Sex-Linked Inheritance
Examples

- **Duchenne Muscular Dystrophy**—X-linked, occurs in about one in 3,600 male births
- Symptoms: waddling gate, toe walking, frequent falls, and difficulty in rising may appear as soon as the child starts to walk
- Muscle weakness intensifies until the individual is confined to a wheelchair
- Death usually occurs by the age of 20

Sex-Linked Inheritance

Examples

- Hemophilia
  - One in 10,000 males
  - X-linked
  - A person's blood does not clot, due to the absence of a particular clotting agent

Sex-Influenced Inheritance

The phenotype is determined by autosomal genes that are expressed differently in males and females

- **Pattern Baldness**—Allele for baldness acts as a dominant in males but acts like a recessive in females
- **Index Finger Length**—An allele for the index finger being longer than the fourth finger acts like a dominant in females and a recessive in males

Linked Genes

All of the genes on one chromosome form a linkage group because they are inherited together

- Crossing-over changes linkage patterns
- Linkage patterns make it possible to map the locations of genes on a chromosome
- The more often crossing-over takes place between two genes the farther apart they must be
- Comparisons of these crossing-over ratios gives us a map of the gene loci

Crossing-Over

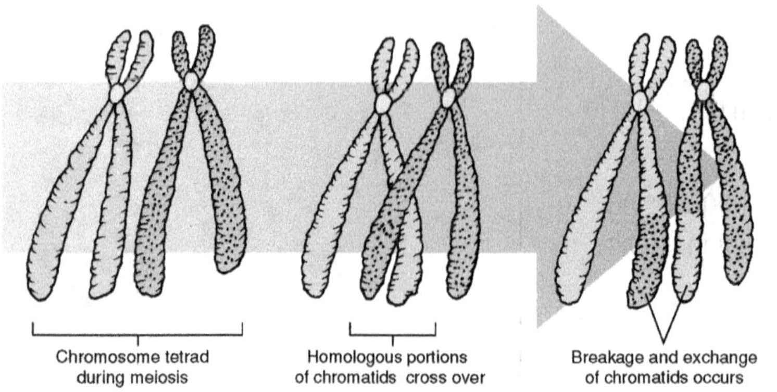

Molecular Basis of Inheritance
DNA Structure and Replication

- Structure was published by Watson and Crick in 1950s
- Made up of nucleotides
  - Sugar (deoxyribose)
  - Nitrogenous base
  - Phosphate
- Double helix

Structure of DNA
Base Pairing

Purines	Pyrimidines
Adenine (A)	Thymine (T)
Guanine (G)	Cytosine (C)

- Weak Hydrogen bonds hold the complimentary bases together
- These make up the rungs of the ladder

Structure of DNA

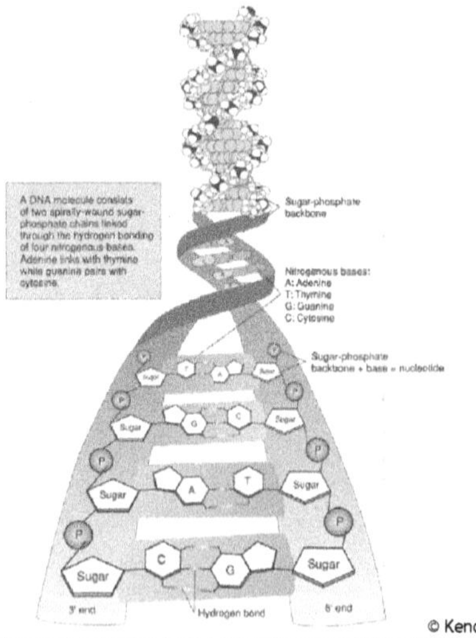

DNA Replication
Unwinding

- The weak hydrogen bonds between the bases break
- The two strands that make up the DNA unwind or unzip
- Caused by a special enzyme called helicase

DNA Replication
Complementary Base Pairing

- New complementary nucleotides are always present in the nucleus
- These fit into place by complementary base pairing
- A-T and C-G

DNA Replication
Joining

- The complementary nucleotides join to form new strands
- This step is carried out by an enzyme called DNA polymerase
- DNA replication is considered semi-conservative because each new double helix is made up of one new strand and one old strand

DNA Replication

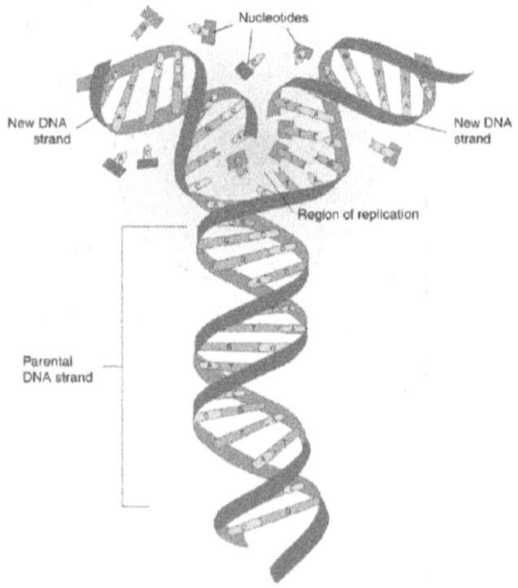

RNA Structure

- Single stranded
- No helix
- Made up of nucleotides
  - Sugar (ribose)
  - Phosphate
  - Nitrogenous base
  - Uracil replaces the thymine in DNA

Types of RNA

Messenger RNA (mRNA)

- Takes a message from DNA in the nucleus to the ribosomes in the cytoplasm
- Each three-base sequence is called a codon
- Produced in the nucleus through the process called transcription
    - DNA segment unzips and mRNA is produced one nucleotide at a time using the DNA template

Transcription

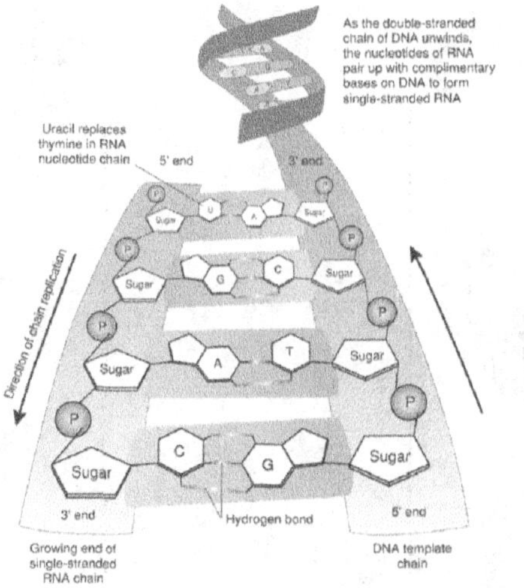

Types of RNA
Ribosomal RNA (rRNA)

- Called structural RNA because, along with protein, it makes up the ribosomes
- A segment of mRNA when it is in physical contact with the ribosome
- Directly associated with the synthesis of polypeptides (proteins)

Types of RNA
Transfer RNA (tRNA)

- Bring amino acids to the ribosomes so that they can be combined to make proteins (translation)
- There is at least one tRNA molecule for each of the twenty amino acids found in proteins
- At one end of the molecule is a group of three bases called an anticodon, which is complementary to the codon of mRNA

Translation

Steps of Translation

- **Initiation**—a small ribosomal subunit attaches to the mRNA near the start codon (AUG)
- **Elongation**—the polypeptide lengthens one amino acid at a time
- **Termination**—protein synthesis comes to an end when a stop codon is reached, the ribosome breaks up and falls off the mRNA molecule

Codon Table

First Base	Second Base: U	Second Base: C	Second Base: A	Second Base: G	Third Base
U	UUU phenylalanine	UCU serine	UAU tyrosine	UGU cysteine	U
U	UUC phenylalanine	UCC serine	UAC tyrosine	UGC cysteine	C
U	UUA leucine	UCA serine	UAA stop	UGA stop	A
U	UUG leucine	UCG serine	UAG stop	UGG tryptophan	G
C	CUU leucine	CCU proline	CAU histidine	CGU arginine	U
C	CUC leucine	CCC proline	CAC histidine	CGC arginine	C
C	CUA leucine	CCA proline	CAA glutamine	CGA arginine	A
C	CUG leucine	CCG proline	CAG glutamine	CGG arginine	G
A	AUU isoleucine	ACU threonine	AAU asparagine	AGU serine	U
A	AUC isoleucine	ACC threonine	AAC asparagine	AGC serine	C
A	AUA isoleucine	ACA threonine	AAA lysine	AGA arginine	A
A	AUG(start) methionine	ACG threonine	AAG lysine	AGG arginine	G
G	GUU valine	GCU alanine	GAU aspartate	GGU glycine	U
G	GUC valine	GCC alanine	GAC aspartate	GGC glycine	C
G	GUA valine	GCA alanine	GAA glutamate	GGA glycine	A
G	GUG valine	GCG alanine	GAG glutamate	GGG glycine	G

Translation

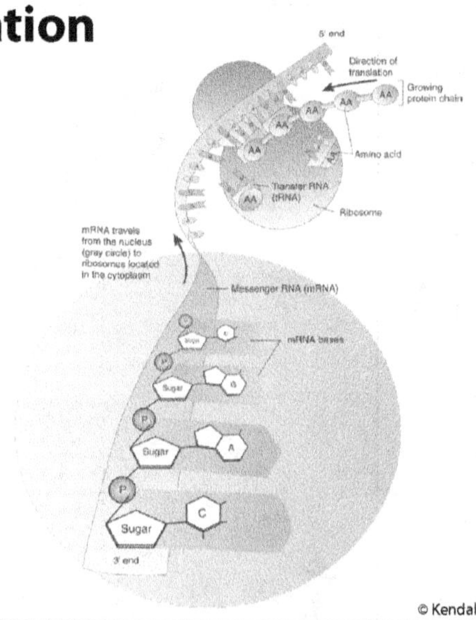

Overview of Translation and Transcription

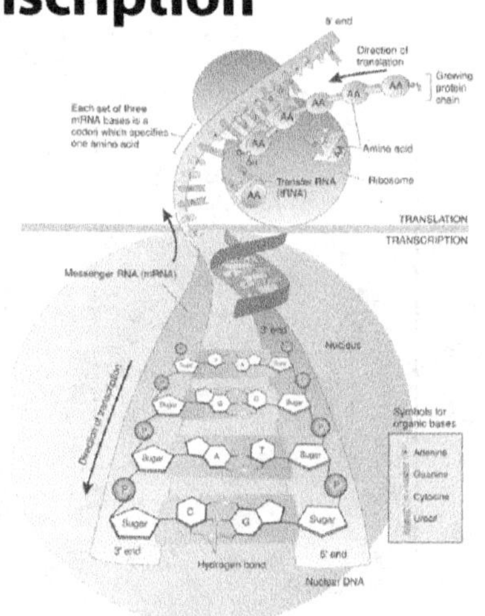

Gene Mutations

Frame Shift Mutations

- The term *Reading frame* applies to the sequence of codons because they are read from a particular starting point
- Occur because of the deletion or addition of a nucleotide from the DNA segment being read
- Results in a completely nonfunctional protein because the sequence of codons is altered

Gene Mutations

Frame Shift Mutations

- Example:
  - THE CAT ATE THE RAT becomes
  - THE ATA TET HER AT with the deletion of the "C"
  - CTH ECA TAT ETH ERA with the addition of a "C"
- Both changes make the sentence nonsense

Gene Mutations

Point Mutations

- Involves a change in a single nucleotide and therefore a change in a single codon
- **Nonsense**—codes for no protein or a stop codon
- **Missense**—codes for the wrong protein
- **Silent Mutation**—substitution that results in the correct amino acid and protein

Unit Five

Evolution . . . no way; I thought
they had disproved that.
How did I get here?

Who am I?

Darwin who?

What is that smell?
What is that smell?
What is that smell?

Did I do that?

Where did I come from?

Who came first,
the chicken or *the egg?*

Evolution

EXAM V

Evidence
Population Genetics
Human
Population Ecology

Evolution
Fossil Evidence

- **Evolution**—the sum of all of the changes that have taken place in living things since the beginning of life
- **Fossils**—the remains and traces of past life
- Traces include:
  - Trails
  - Footprints
  - Burrows
  - Worm casts
  - Preserved droppings

Archaeopteryx
Transitional Fossils

- Link between reptiles and birds
  - Scales on legs and feet
  - Egg layers
  - No teeth
- Had feathers and wing claws
- Most likely a poor flyer

Archaeopteryx

Geologic Time Scale
Relative Dating Method

- **Law of Superposition**—fossils are found in sedimentary rock which is laid down over long periods of time in layers (strata)
- Since the top layers are younger than those layers that are under them
- It can be estimated that the fossils found in these younger layers are also younger
- Does not give an age in years, but in relation to other fossils

Geologic Time Scale
Absolute Dating

- Relies on radioactive dating techniques to assign an actual date to a fossil
- All radioactive isotopes have a specific half-life
  - **Half-life**—the length of time it takes for one half of the sample to change into another stable element
- Carbon 14 is the only radioactive isotope found in organic matter
- **Very reliable up to 50,000 years**

Geologic Time Scale
Absolute Dating

- Certain radioactive isotopes can be used to date rocks directly, the age of the fossil in the rock can be inferred
- The ratio of potassium 40 to argon 40 trapped in rock is often used
- **Used in rock less than 100,000,000 years old**

Geologic Time Scale
Absolute Dating

- Ratio of isotope Uranium 238 to lead 207
- Has a very long half-life
- **Used in rocks older than 100,000,000 years**

Mass Extinctions
Theories

- **Extinction**—the death of every member of a species
- **Mass Extinctions**—times when a large percentage of the existing species become extinct in a relatively short time
- There have been at least five mass extinctions
- Occurred about every 26 million years

Mass Extinctions
Cretaceous Extinction

- The most well known
- The extinction of the dinosaurs
- Due to an asteroid that exploded, producing meteorites that fell to Earth
- A large meteorite striking the earth could have produced a cloud of dust that blocked out the sun's warming light

Mass Extinctions

Cretaceous Extinction

- First plants died, then the largest herbivores, then the largest carnivores may have died in the cold
- Some believe that less catastrophic events killed the dinosaur
- Possibly a virus
- Competition for space and resources caused overcrowding

Biogeographical Evidence

- **Biogeography**—the study of the distribution of plants and animals throughout the world
- Related forms evolve in one location and spread out into other regions
  - There are no rabbits in South America—they evolved someplace else and had no means of getting there
  - There are no snakes found on certain islands for the same reason

Biogeographical Evidence
Continental Drift

- The position of the continents is not fixed
- They are constantly moving (floating) on a sea of molten rock called magma
- 225 million years ago—Pangea
- 135 million years ago—Laurasia and Gondwanaland
- 65 million years ago—present continents separate

Pangea

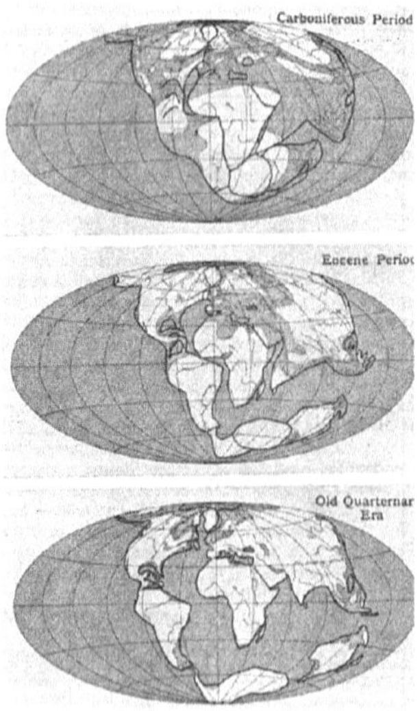

Anatomical Evidence
Homologous Structures

- Structures that are similar because they were inherited from a common ancestor
- **Example:**
  - Forelimbs of mammals
- The wing of a bird and an insect are **Analogous Structures** (they are adapted for flying, but are structurally unrelated)

Animal Homology

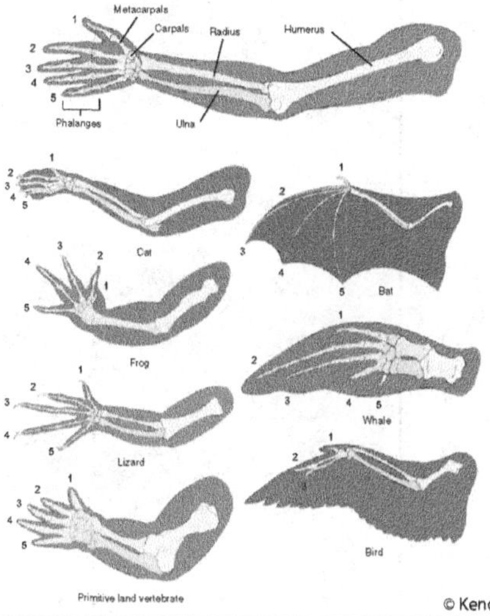

Anatomical Evidence
Homologous Structures

- **Ernst Haeckel** (1834–1919)
- "Ontogeny recapitulates phylogeny"
- The embryonic development of a species restates the many stages of evolution the species has gone through

Biochemical Evidence

- All organisms have certain biochemical molecules in common
- The degree of similarity between DNA and amino acid sequences is thought to indicate the degree of relatedness

Population Genetics

- **Population**—all members of a single species occupying a particular area at the same time
- **Gene Pool**—the various alleles of all the gene loci in all the members of a population
- **Hardy-Weinberg Law**—allele frequencies in a sexually reproducing population come to an equilibrium that is maintained generation after generation
  - No mutations
  - Random mating
  - No gene flow
  - No natural selection

Population Genetics
Hardy-Weinberg Law

- In real life these conditions rarely if ever occur
- Allele frequencies do change
- The significance is that it tells us what factors cause evolution—those that violate the conditions listed

Hardy-Weinberg Law

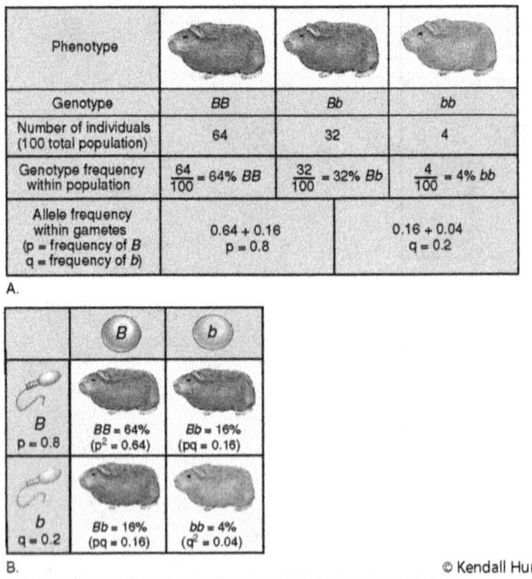

Agents of Evolutionary Change

Mutations

- Provide new alleles
- Underlie all other mechanisms that produce variation
- The raw material for evolutionary change
- The adaptive value of a mutation is entirely dependent on environmental conditions

Agents of Evolutionary Change

Genetic Drift

- Refers to the change in allele frequencies of a gene pool due to chance
- **Founder Effect**—occurs when only a few individuals found a colony and only a fraction of the total genetic diversity of the original pool is represented
- **Example:**
  - Pennsylvania Amish

Agents of Evolutionary Change

Gene Flow

- The movement of alleles between populations
- **Emigration**—the loss of genes in a gene pool by an individual leaving a population
- **Immigration**—the gaining of alleles in a gene pool by an individual joining a population

Agents of Evolutionary Change
Nonrandom Mating

- Occurs when individuals pair up according to their genotypes and phenotypes, **not by chance**

Agents of Evolutionary Change
Natural Selection

- Based in part on the work of Charles Darwin
- The process by which populations become adapted to their environment
- "Survival of the fittest"
    - Fitness is measured by how reproductively successful its offspring are in the next generation

Agents of Evolutionary Change

Natural Selection

- **Variation**—the members of a population differ from one another
- **Inheritance**—many of these differences are heritable genetic differences
- **Differential Adaptedness**—some of these differences affect how well an organism is adapted to its environment
- **Differential Reproduction**—individuals that are better adapted are more likely to reproduce, and their offspring will make up a greater proportion of the next generation

Speciation

A group of interbreeding subpopulations that share a gene pool and that are isolated reproductively from other populations

- **Premating**
  - Habitat (occupy different habitat)
  - Temporal (reproduce at different time of year or day)
  - Behavioral (courtship behaviors differ)
  - Mechanical (genitalia unsuitable for one another)
- **Postmating**
  - Gamete isolation (sperm cannot reach egg)
  - Zygote mortality (zygote does not survive)
  - Hybrid sterility (hybrid survives but is sterile)
  - F_2 fitness (hybrid survives but has reduced fitness)

Modes of Evolutionary Change

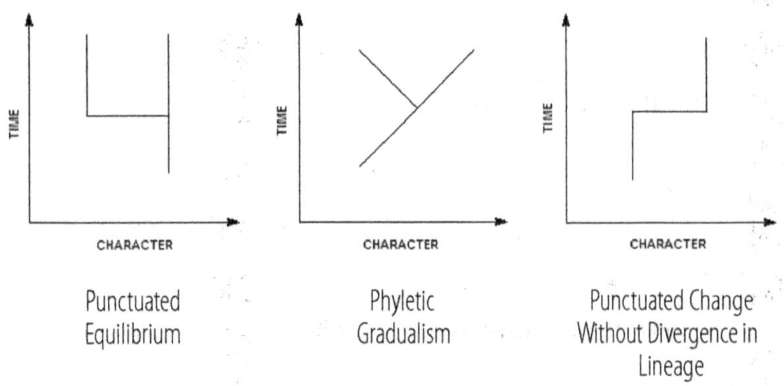

Human Evolution

Early Primates

- **Hominoid Superfamily**—group contains species that would later develop into the hominid line and the great apes
- ***Aegyptopithecus***—found in the later Oligocene, 28–29 MYA, in Egypt, and having a somewhat developed forehead, small incisors, lower premolars, and large molars
- **Proconsul**—three species; major, africanus, and nyanzae, all showing hominid characteristics
- ***Kenyapithecus wickeri***—a possible forerunner for the pongid and hominid lines

Hominids

Climatic Cooling

- Planet-wide cooling and drying during the Late Miocene (6–5.3 MYA) probably triggered speciation of hominoid super-family into the hominid family
- Giant forests gave way to vast grasslands
- Tree dwellers were replaced by the better adapted upright posture of more modern man:
    - Better visibility
    - Faster
    - Safer
    - Hands-free locomotion

Hominid Evolutionary Tree

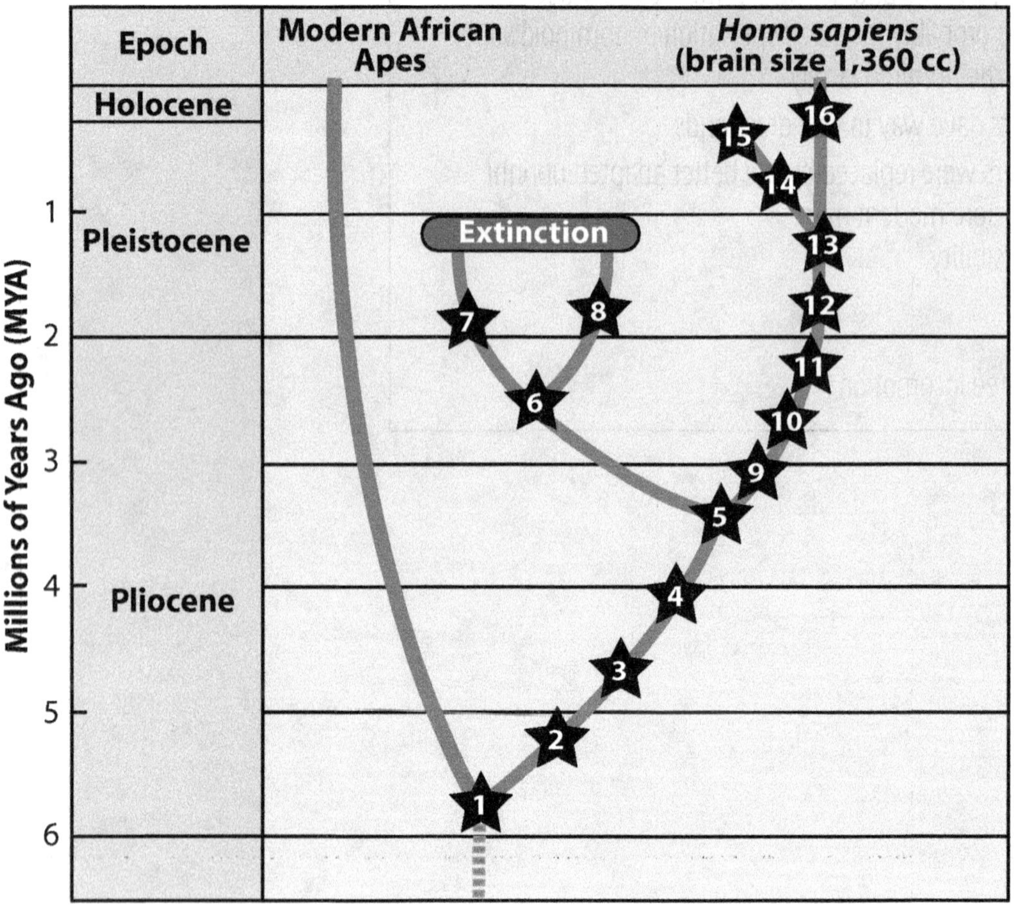

- 8 *P. robustus*
- 7 *P. boisei*
- 6 *Paranthropus Aethiopicus*
- 5 *A. africanus*
- 4 *A. afarensis*
- 3 *Australopithecus anamensis*
- 2 *Ardipithecus (Australopithecus) ramidus*
- 1 *Kenyapithecus wickeri*

- 16 *H. sapien*
- 15 *H. neanderthalensis*
- 14 *H. heidelbergensis*
- 13 *H. antecessor*
- 12 *H. erectus*
- 11 *H. ergaster*
- 10 *Homo habilis*
- 9 *H. rudolfensis*

Adapted from Sylvia Mader. "Inquiry into Life." Figure 31.16; pg. 661.

Hominids
Ardipithecus (Australopithicus) ramidus

- Ethiopia
- 4.4 MYA
- Hominid teeth
- Upright posture
- Semi-bipedal mode of locomotion
- Temperate or tropical forest

Hominids
Australopithecines

- General information:
  - Found only in Africa
  - Genus is 5 million years old
  - Human-like jaws and teeth
  - Ape-like skull (small)
  - Gracile (small and delicate)
  - Bipedal (but, still tree climbers)
  - Sexual dimorphism

Skeletal Changes

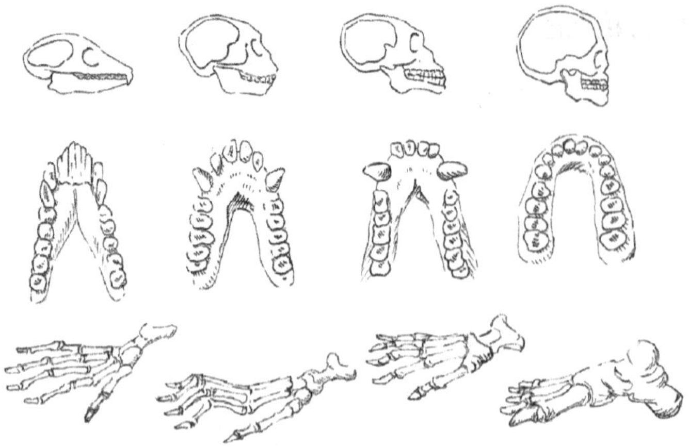

Image courtesy of Kevin Mercer

Australopithecines

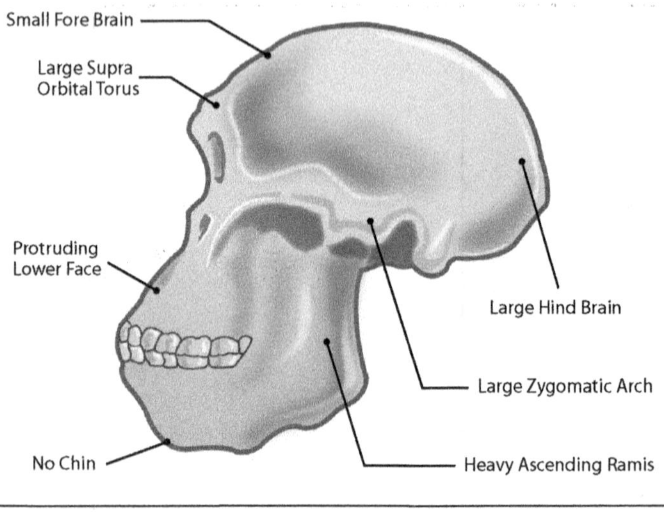

A. anamensis

- 4.2 to 3.5 MYA
- East Africa only
- Forest dweller

A. afarensis

- 5 to 3 MYA
- 400 cc braincase
- Arms longer than humans but, shorter than apes
- About 4 feet tall
- Lived in forests and savannas
- Tree climbing abilities
- Often found in groups
- Ape-like (grasping) big toe

A. africanus

- 3 to 1 MYA
- Gracile variety
- 440 to 485 cc braincase capacity
- Totally bipedal
- More human-like big toe

Paranthropus
General Information

- Formerly classified as a robust form of *Australopithecus*
- Massive teeth and jaws indicate vegetarian diet
- Could have produced and manipulated simple tools
- Tropical forest dweller

Paranthropus

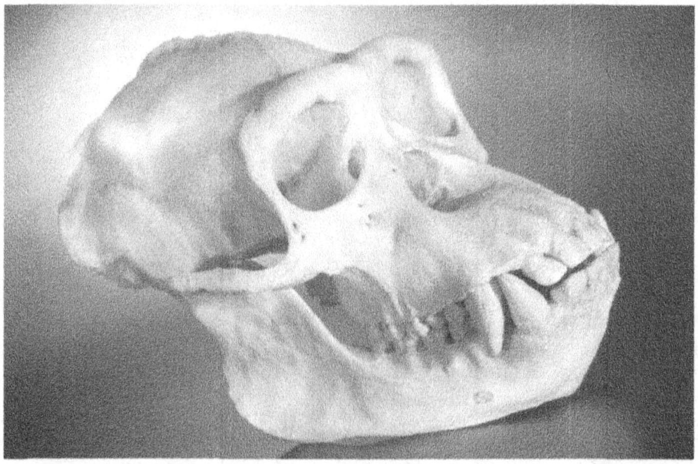

© JupterImages, Inc.

Paranthropus Aethiopicus

- 3.3 to 2.6 MYA
- Suspected progenitor of the genus
- The "Black Skull" of Tanzania

P. boisei

- 2.6 to 1.2 MYA
- Largest variety
- Some with crests and flanges
- Smallest hominid braincase (410 cc)

P. robustus

- 2.6 to 1.8 MYA
- "Zinjanthropus" or "Nutcracker Man"
- 530 cc braincase capacity

Early Homo

General Characteristics

- Much larger and complex brains than *Australopithecines*
- Significantly increased tool making ability
- Less prominent cranial ridges and no flanges
- Less protruding face
- Emergence of third molar (wisdom tooth)
- Probably pair-bonded
- Omnivorous scavengers

Variations Found in Genus *Homo*

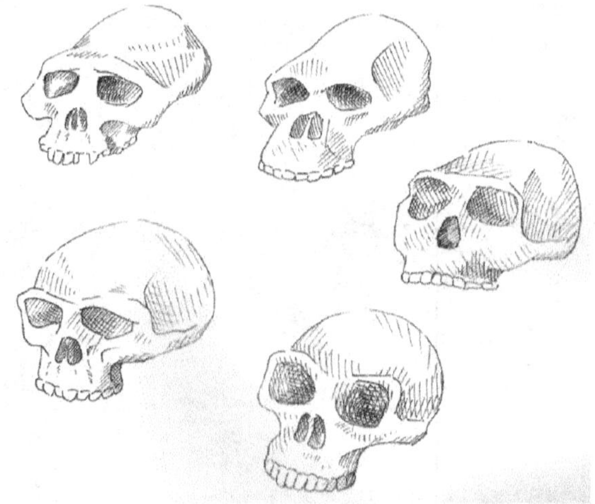

Image courtesy of Kevin Mercer

Evolution of Hominid
Brain Size and Shape

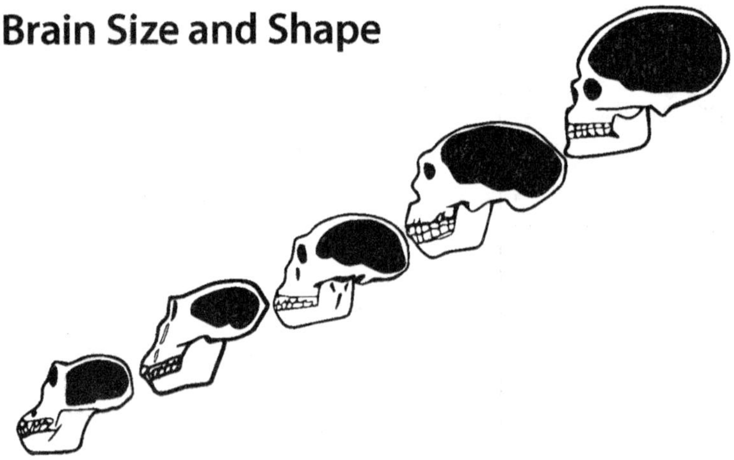

Image © Hano Uzeirbegovic, 2009.
Used under license from Shutterstock, Inc.

Homo Rudolfensis

- 2.5 to 1.8 MYA
- Simple tool technology
- Catchment scavenging
- Congregated near locations that offer many different needed resources (lake margins, stream confluences, and rock outcroppings)
- 750 cc braincase capacity

H. habilis

- 2 to 1.6 MYA
- Omnivorous teeth and third molars (like moderns)
- Rounder head with less pronounced brow ridges
- Slight chin and smaller teeth
- Almost hairless face
- Stood 3.7 to 4.2 feet tall
- May have had the first language (speech)
- More advanced tool use
- May have produced first architectural structure (Stone circle)

H. habilis

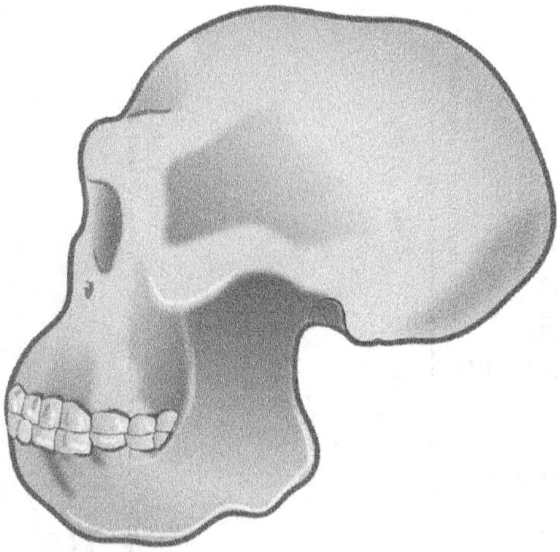

H. Ergaster

- Much larger brain than *habilis* (1000 cc)
- Less protruding, lighter facial features
- Territorial scavenging (exploiting many locals across a large area)
- Wide, sparse landscapes of the plains and temperate, drier climates
- In the Middle Pliocene some migrated to the Eurasian continent
- Much more complex tools
- First true hunter, as opposed to scavenger

H. Erectus

- 2 MYA to 400 YA
- Larger brain and body
- Fully upright and bipedal
- Longer skull, lower forehead, thicker cranial bones, larger projecting brow ridges, and a heavier face than *H. sapiens*
- As tall as modern humans
- Found in tropical Africa, but mainly in Asia, Indonesia, and points in between
- Cooling due to glaciation in Middle Miocene may have prompted them to follow migrating herds of bovine out of Africa into Asia

H. Erectus

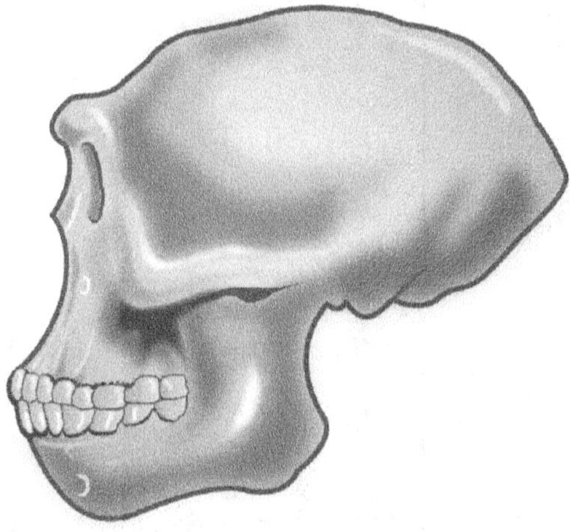

H. Antecessor

- 1 MYA to 800 YA
- Relatively modern face
- Lived in Europe and North Africa
- Oak, pine, and beech forest
- Closest common ancestor to *sapiens* and *heidelbergensis*
  - Perhaps arose in Africa and migrated to Europe, spawning *heidelbergensis*, while relatives left behind gave rise to *sapiens* in Africa

H. heidelbergensis

- 600,000 YA
- Taller than modern humans
- Precursors to Neanderthals
- Robust physiques, big noses, no chins, and heavy brows
- Capable of "group hunting" large game
- Hints of knowledge of mortality (ritualistic burials)
- Scrape marks on bones may be evidence of defleshing (cannibalism?)

H. neanderthalensis

- 230,000 to 29,000 YA
- Larger bodies, more barrel-chested than sapiens
- Brain larger than modern humans
- Lack of frontal lobe and characterized by an occipital bun
- Ranging from Europe to Middle East (during the Ice-age) as climate warmed they were also found in Asia and North Africa
- Used wooden spears for stabbing and thrusting (never throwing)
- Presence of jewelry indicates attempts by moderns to trade with Neanderthals
- A flute was discovered (43 to 67 YA) matches the seven-note diatonic scale of modern human music

H. Sapiens
Early Moderns

- 200,000 to ? YA
- Largest braincase relative to size (1200–1400 cc)
- Capable of lanquage, symbolism, and other abstract concepts
- Slighter body than *neandertalensis*
- Evolved in Africa

H. Sapiens
Cro-Magnon

- 50,000 YA to present
- Anatomically similar to modern humans
- Adapted to all extremes of heat and cold
- Took advantage of lower sea levels to cross land bridge to North America (20 to 25,000 YA)

Cro-Magnon

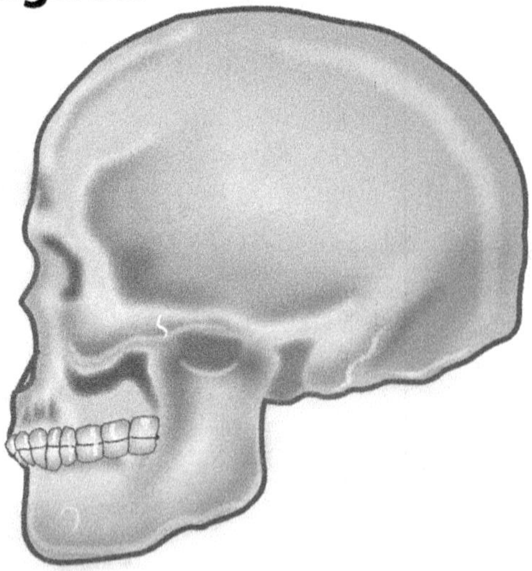

Cro-Magnon

Image courtesy of Kevin Mercer

H. Sapiens
Cro-Magnon

- Invented the bow and arrow and animal traps
- More sophisticated spear points
- Began marking time with lunar phases
- People lived in tents and other man-made shelters in groups of several families

H. Sapiens
Cro-Magnon

- Nomadic hunter-gatherers
- Had elaborate rituals for hunting, birth, and death
- First to domesticate animals
- First to leave extensive works of art

Population Interactions

- Since archaic species existed at the same time as *H. sapien sapien* and *H. sapien neandertalensis* . . .
  - Did they interbreed?
  - Were they responsible for blending the gene pool of all three?
    - Creating a "hybrid" species that replaced all three?

Multi-Regional Continuity Model

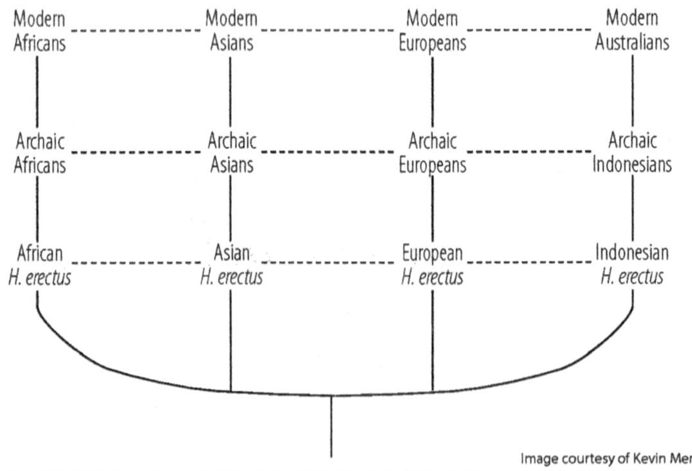

Image courtesy of Kevin Mercer

Population Ecology

Scope of Ecology

- **Ecology**—the study of interactions of organisms with each other and their physical environment
- **Population**—all the organisms within an area belonging to the same species
- **Community**—consists of all the various populations interacting at a locale
- **Ecosystem**—contains a community of populations and also the abiotic (nonliving) environment
- **Biosphere**—portion of the entire earth's surface where life exists

Density and Distribution

Population Density

- The number of individuals per unit area
  - Uniform—evenly distributed in an area
  - Random—randomly located within an area
  - Clumped—high density in some areas and low density in others (the most common type)
- Example:
  - Humans

Characteristics of Populations
Patterns of Population Growth

- Per capita rate of increase:

$$\frac{\text{\# of births per year} - \text{\# of deaths per year}}{\text{\# of individuals in original population}}$$

- **Biotic Potential**—the highest possible per capita rate of increase for a population

Biotic Potential
Controlled by Four Factors

1. Usual number of offspring per reproduction
2. Chances of survival until age of reproduction
3. How often each individual reproduces
4. Age at which reproduction begins

Exponential Growth

J-Shaped Growth Curve

- Lag Phase—growth is slow because population is small
- Exponential Growth Phase—growth is accelerating and the population is exhibiting its biotic potential

J-Shaped Growth Curve

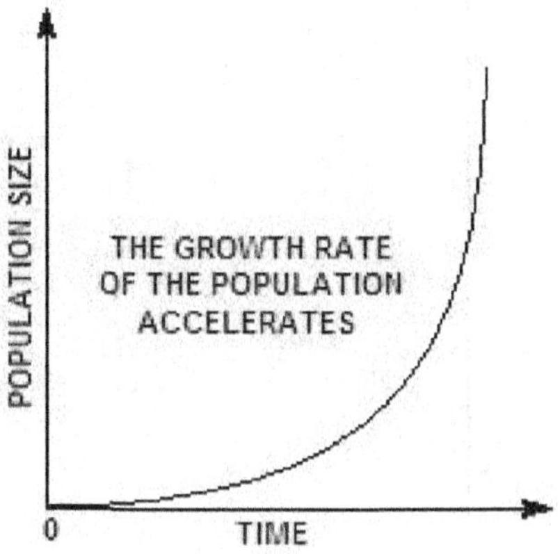

Image courtesy of Kevin Mercer

Exponential Growth
Environmental Resistance

- All those environmental conditions that prevent a population from achieving their biotic potential
- Example:
  - Limited supply of food
  - An accumulation of waste products
  - Increased competition
  - Increased predation

Logistic Growth
S-Shaped Growth Curve

- **Lag Phase**—growth is slow because the population is small
- **Exponential Growth Phase**—growth is accelerating due to biotic potential
- **Deceleration Phase**—the rate of population growth slows down
- **Stable Equilibrium Phase**—there is little if any growth because births and deaths are about equal

S-Shaped Growth Curve

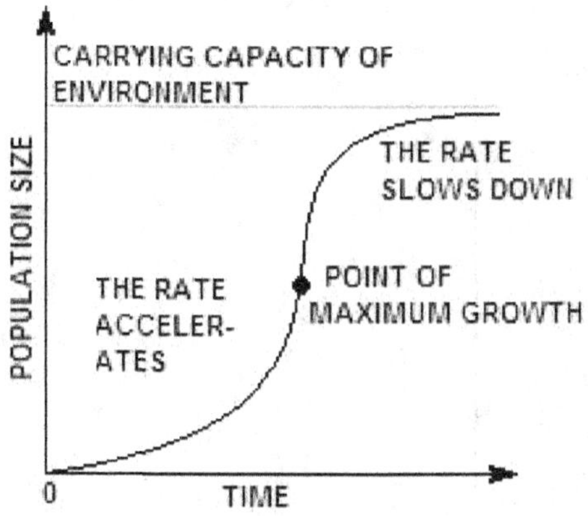

Image courtesy of Kevin Mercer

Survivorship

- **Cohort**—an original group of individuals born at the same time
- **Survivorship Curve**—graphical representation of those individuals that are still alive after a given amount of time

Survivorship Curves
Three Types

- **Type I**—most individuals survive past the midpoint, and death does not come until near the end of the life span (humans)
- **Type II**—survivorship decrease at a constant rate throughout the life span
- **Type III**—most individuals die very young

Survivorship Curves

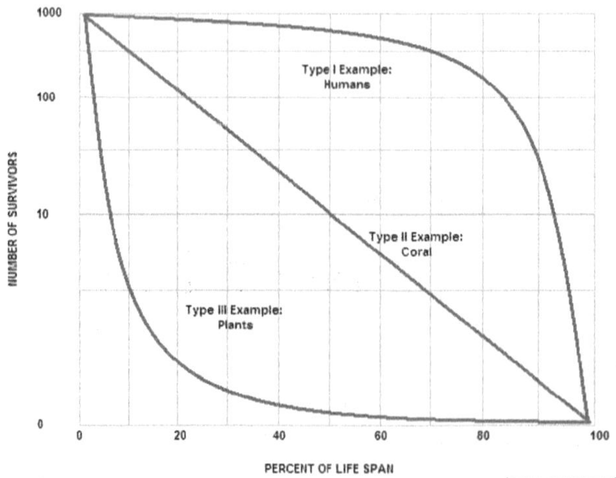

Image courtesy of Kevin Mercer

Regulation of Population Size
Density-Dependent Effect

- The number of individuals in a given area regulates the number of individuals that can exist in that area
- The more successful your generation . . . the less successful the next generation can be

Human Population Growth
Exponential Growth

- **Doubling Time**—the length of time it takes a population to double in size
- The current doubling time is 47 years
  - You are expected to be the last generation to see a decrease in doubling time

Human Population Growth
History and Predictions

- The **first** billion did not occur until 1800
- The **second** billion occurred in 1930
- The **third** billion occurred in 1960
- Today there are over **6.5 billion**

Human Population Growth
History and Predictions

- The world's population may level off at 8, 10.5, or 14.2 billion
    - 8 billion = Global vegetarianism
    - 10.5 billion = Global vegetarianism and ban on all domesticated animals, including pets
    - 14.2 billion = Global vegetarianism, ban on domesticated animals, and no fossil fuel use
- Depending on the rate at which the per capita increase declines

Human Population Growth

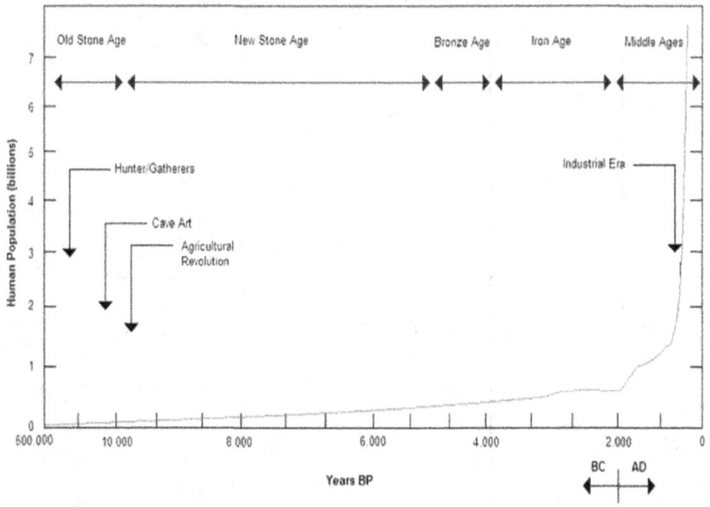

Image courtesy of Kevin Mercer

Human Population Growth
Age Distributions

- More-Developed Countries (MDCs)—typified by countries in North America and Europe
  - Those in which population growth is slow and most people enjoy a good standard of living
- Less-Developed Countries (LDCs)—countries in Latin America, Africa and Asia
  - Where population growth is rapid and standard of living is typically very low

MDC–Age Distributions

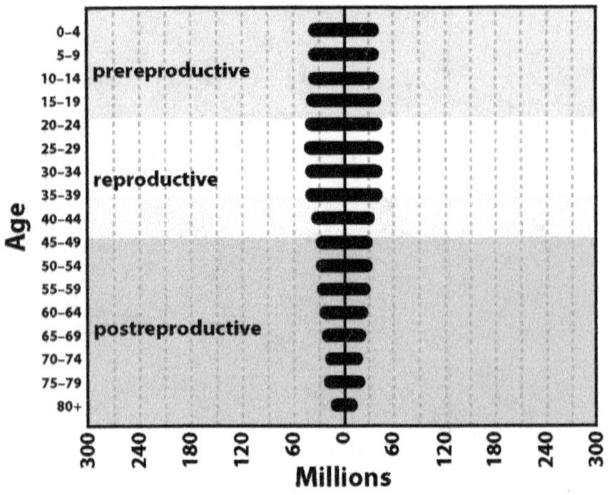

Adapted from Sylvia Mader. "Inquiry into Life." Figure 33.10; pg. 696.

LDC–Age Distributions

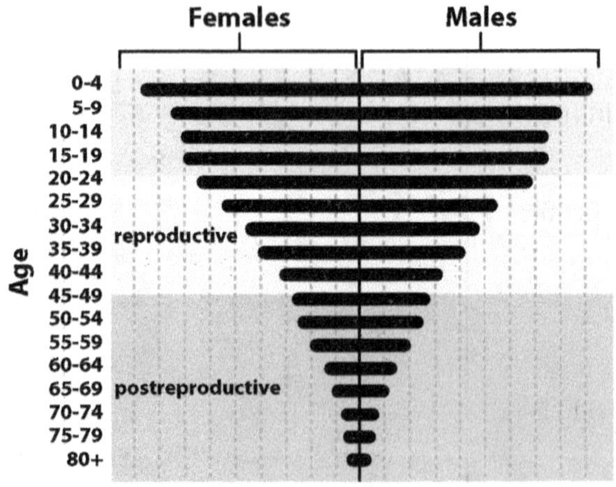

Adapted from Sylvia Mader. "Inquiry into Life." Figure 33.10; pg. 696.

Human Population Growth
A Sustainable World

- The growing populations in LDCs are putting a strain on the environment
- Excessive consumption of resources by the MDCs are also stressing the environment
- Development will meet economic needs of all peoples while protecting the environment for future generations

Carbon Dioxide Emissions

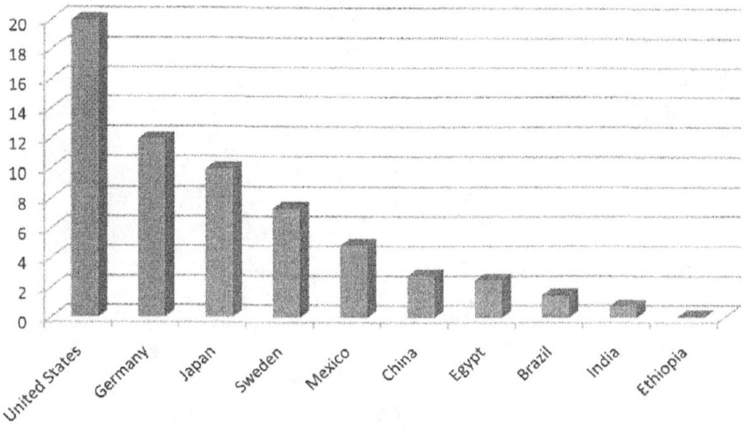

Image courtesy of Kevin Mercer

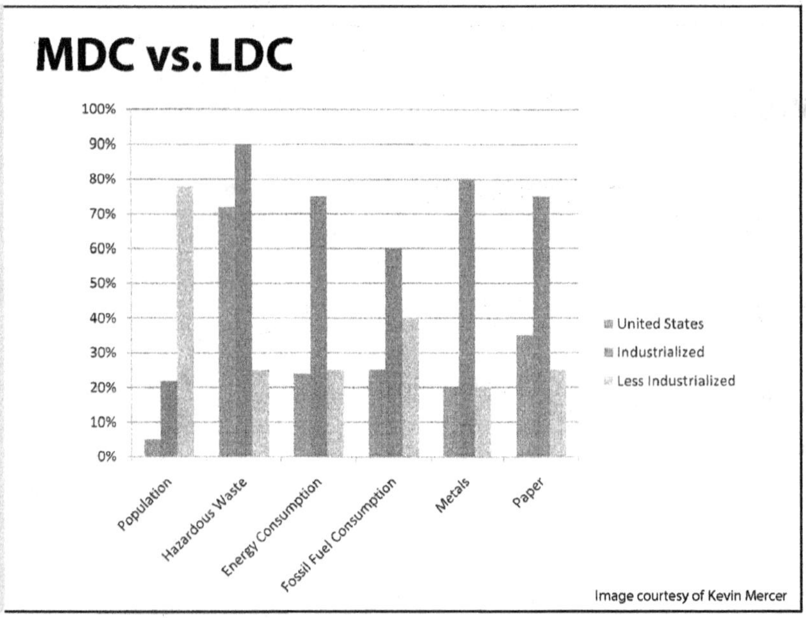
MDC vs. LDC
100%
90%
80%
70%
60%
50%
40%
30%
20%
10%
0%
Population
Hazardous Waste
Energy Consumption
Fossil Fuel Consumption
Metals
Paper
United States
Industrialized
Less Industrialized
Image courtesy of Kevin Mercer

Unit Six

Who are we and how can we
do the least amount of harm to
"Mother Earth" as we *pass*
through this *life?*

Is there any hope
for our survival?

What is that smell?

**Why do we do
the things we do?**

Are we really doomed?

Ernst Heinrich Philipp August Haeckel?
No way, not him again!

Ecology

EXAM VI

Community Ecology
Animal Behavior
Terrestrial Biomes
Aquatic Biomes

Community Ecology
Composition

- A listing of the various species in the community
- **Example:**

Coniferous Forest	Tropical Rain Forest
Moose	Monkey
Lynx	Sloth
Snowshoe Hare	Anteater
Bear	Kinkajou
Red Fox	Jaguar
Wolf	Tapir

Diversity

- Includes both species richness (the number of species) and evenness (the relative abundance of individuals of different species) of a community
- The greater the diversity the greater the number and the more even the distribution of species

Habitat

- Where an organism lives and reproduces in an environment
- **Example:**
  - Forest floor
  - Swift stream
  - Ocean's edge

Ecological Niche

- The role an organism plays in its community, including its habitat and its interactions with other organisms
- Aspects of a niche:

Animal	Plant
Time of day for feeding and time of year for reproduction	Seasons of year for growth and reproduction
Habitat and food requirements	Sunlight, soil, and water requirements
Relationships with other organisms	Relationships with other organisms
Effect on abiotic environment	Effect on abiotic environment

Competition

- Occurs when members of different species try to utilize a resource (light, space, or nutrients) that is in limited supply
- **Competitive Exclusion Principle**—no two species can occupy the same niche at the same time, if resources are limiting
- **Resource Partitioning**—partitioning the available resource so as to avoid direct competition with each other
    - Subdividing the niche

Predation

- Occurs when one living organism, the predator, feeds on another, the prey

Prey Defenses

Mimicry

- **Batesian Mimics**—occur when one species resembles another that possesses an antipredator defense
  - It can help a predator capture food or help prey avoid capture
    - Monarch butterfly and Viceroy butterfly

Batesian Mimics

Monarch

Image © Varina and Jay Patel, 2009. Used under license from Shutterstock, Inc.

Viceroy

© JupiterImages, Inc.

Prey Defenses
Mimicry

- **Müllerian Mimics**—several unrelated but protected species come to resemble each other
- **Example:**
  Yellow Jacket, Masarid Wasp, Sand Wasp, Anthidiine Bee
  - Share the yellow and black color pattern
  - Thus reinforcing the message that these colors are to be avoided

Müllerian Mimics

"Red next to Black is a friend of Jack" . . . Scarlet King Snake.

"Red next to Yellow will kill a fellow" . . . Eastern Coral Snake.

Prey Defenses

- Camouflage
- Fright
- Flocking together
- Warning coloration

Symbiosis
Parasitism

- Symbiotic relationship in which the parasite derives nourishment from another called the host
- Host organisms also provide:
  - A place to live
  - A place to reproduce
  - A mechanism for dispersing offspring to new hosts

Symbiosis
Commensalism

- A symbiotic relationship between two species in which one species is benefited and the other is neither benefited nor harmed
- **Example:**
  - Remoras and sharks
  - Clownfish and anemones

Symbiosis

Mutualism

- A symbiotic relationship in which both members of the association benefit
- Example:
  - Lichen = fungi + algae

Ecological Succession

- A change in a community composition over time
- **Primary Succession**—occurs in areas where there is no soil formation, such as after a volcanic eruption or glacial retreat
- **Secondary Succession**—occurs in areas where soil is present, such as when a cultivated field returns to its natural state

Ecological Succession
Models of Succession

- **Climax-Pattern Model**—says that particular areas will always lead to the same type of community, which is called a *"Climax Community"*
- Climate is one of the determining factors of which plant community will be established
- Others include:
  - Altitude
  - Soil type
  - Animal interaction

Ecological Succession
Models of Succession

- **Facilitation Model**—each stage in succession facilitates the next stage
- **Example:**
  - Shrubs can not grow on sand dunes until dune grasses have made the soil suitable for them

Ecological Succession
Models of Succession

- **Inhibition Model**—states that colonists hold on to their space and inhibit the growth of other plants until the colonists die or are damaged
- **Tolerance Model**—states that different kinds of plants can colonize an area at the same time

Intermediate Disturbance Hypothesis

- States that a moderate amount of disturbance is required for a high degree of community diversity
- Example:
  - Fire
  - Flood

Island Biogeography

- States that every island has a balance between the rate of immigration of new species and the rate of extinction of species that are already on the island
  - No two species will occupy the same niche
  - The "new" species will not move in until the "old" species is no longer the "fittest"

Animal Behavior

- **Behavior**—any action that can be observed and described
- **Genetic Bases**—the anatomy and physiology of the animal (which is inherited) is suitable for the behavior
  - Genes influence the development of neural and hormonal mechanisms that control behavior

Development of Behavior

- **Learning**—a durable change in behavior brought about by experience
- **Operant Conditioning**—the gradual strengthening of stimulus-response connections **(Alexander Pavlov)**
- **Imprinting**—chicks, ducklings, and goslings follow the first moving object they see after hatching **(Konrad Lorenz)**

Adaptiveness of Behavior

Female Choice

- **Good Genes Hypothesis**—females choose mates on the basis of traits that improve their chances of survival
- **Run-Away Hypothesis**—females choose mates on the basis of traits that make them attractive to females

Male Competition
Dominance Hierarchy

- Higher-ranking individual has greater access to resources than a lower-ranking individual
- Displays begin early in life
- Often start as play in organized games
- This allows physical contact in a relatively "Controlled" environment

Territoriality

A specific territory is defended against competitors

- Vocalizations
- Displays
- Attacks
    - For man to reach this level one or both of the following are typically involved:
        - An audience
        - Alcohol or drugs

Animal Societies

Communicative Behavior

- An action by a sender that influences the behavior of a receiver
- **Chemical Communication**—the advantage of working both night and day
  - Works for a relatively long time
  - Unable to modify the signal.
    - **Pheromone**—chemical signals in low concentration that are passed between members of the same species

Auditory Communication

- Faster than chemical communication
- Night or day
- Works at a distance
- Can be modified by:
  - Volume
  - Duration
  - Pattern
  - Repetition

Visual Communication

- Most often used by species that are active during the day
- Dominant in humans
  - Colors
  - Styles
  - Name brands
  - Logos

Tactile Communication

- Occurs when one animal touches another
- Most intimate form of communication
  - Holding hands
  - Kissing
  - Caressing

Sociobiology

- Applies the principles of evolutionary biology to the study of social behavior in animals

Altruism vs. Self-Interest

- Altruism
  - Behavior that has the potential to decrease the lifetime reproductive success of the altruist
  - While benefiting the reproductive success of another member of the group

Altruism vs. Self-Interest

- Army ants have a single breeding pair; the queen and her mate
- Three different sizes of sterile female workers
  - **Smallest (Nurses)**—take care of the queen and her larvae (feeding them and keeping them clean)
  - **Medium (Workers or Drones)**—go out on raids to collect food (majority of the population)
  - **Largest (Soldiers)**—protect the nest and raiding parties

Natural Selection

- **Direct Selection**—the parent passes the gene directly on to the offspring
- **Indirect Selection**—an animal can help a relative reproduce and thereby pass the gene to the next generation via the relative
- **Inclusive Fitness**—includes both personal reproductive success and that of relatives

Biosphere
Climate

- The distribution of biomes in the biosphere is dependent upon:
  - Variations in reception of solar radiation due to a spherical earth
  - The tilt of the earth's axis as it rotates about the sun
  - Distribution of land masses and oceans
  - Topography (landscape) features

Air Circulation

- The tilt of the earth causes an unequal heating of the earth's surface creating the seasonal changes, except at the equator
- When the Northern Hemisphere is having winter the Southern Hemisphere is having summer
- Heat always passes from warmer areas to cooler areas

Air Circulation

- Because the earth rotates; these moving air masses (winds) are modified into three large circulation cells in each hemisphere
- At the equator the sun heats the air and evaporates the water
- The warm moist air rises and cools, losing much of its moisture as rain
- The greatest amounts of rainfall are near the equator

Air Circulation

- The rising air moves toward the poles
- At 30° north and south latitudes it sinks toward the earth and is reheated
- As the air descends and warms it becomes very dry, creating zones of low rainfall
    - The great deserts of Africa, Australia, and the Americas occur at these latitudes

Air Circulation

- At about 60° north and south latitudes the air once again rises and cools
- This produces another zone of heavy rainfall
  - This moisture supports the great forests of the temperate zones

Distribution of Air Circulation

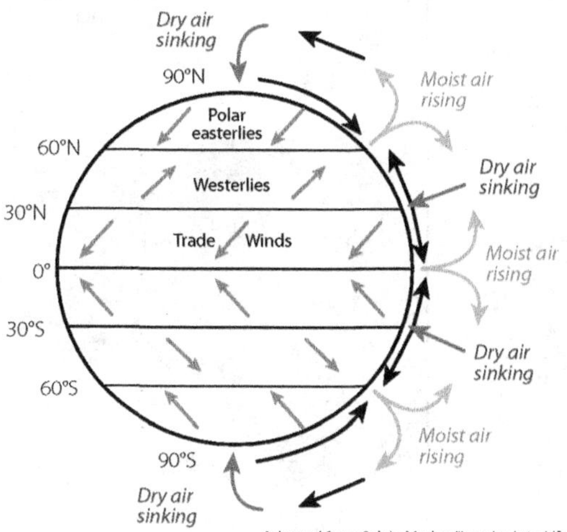

Adapted from Sylvia Mader. "Inquiry into Life." Figure 35.2; pg. 719.

Air Circulation
Topography

- The physical features of the land
- One topographical feature that affects the climate is the presence of mountains
  - As air blows up and over the mountain range it rises and cools
  - One side of the mountain (windward) receives much more rain than the other (leeward) side
- The leeward side of the mountain is said to be in a rain shadow

Rain Shadow

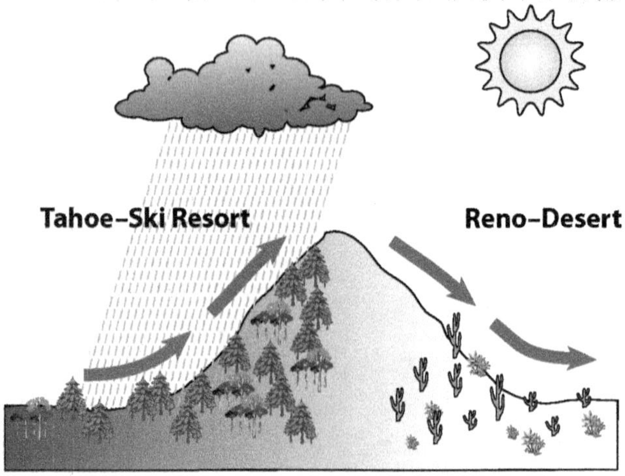

Biomes of the World

- The largest geographical unit of the biosphere
- Has a particular mix of plants and animals that have adapted to living under certain environmental conditions
- Biomes gradually change from one type to another

Terrestrial Biomes

Tundra

- Encircles the earth just south of the ice-covered polar seas in the Northern Hemisphere
  - Covers about 20% of the earth's land surface
  - Cold and dark much of the year
  - Only about 20 cm of rainfall per year
  - Only the topmost layer of earth ever thaws, below that is always frozen (permafrost)
  - No trees because growing season is too short, and permafrost prevents adequate root growth
  - Few animals live in the tundra year round

Terrestrial Biomes

Coniferous Forests

Found in three locations:

1. **Taiga**—extends around the world in the northern part of North America and Eurasia
2. Near mountain tops (Montane Coniferous Forest)
3. Along the Pacific coast of North America as far south as California

Terrestrial Biomes

Coniferous Forests

- **Taiga (Swampland)**—typifies the coniferous forest with its cone-bearing trees such as spruce, fir, and pine that do not lose their leaves seasonally
- Well adapted to the cold with thick coverings on both leaves and bark
- Leaves (needles) are modified to prevent water loss, freezing, and damage due to heavy snow
- Limited understory growth

Terrestrial Biomes
Temperate Deciduous Forest

- Found south of the Taiga in North America, eastern Asia, and much of Europe
- Moderate climate with relatively high rainfall (75–150 cm)
- Well-defined seasons, with a 140 to 300 day growing season
- Trees such as oak, beech, maple, hickory, and others with broad, flat leaves
- Grow leaves in the spring and lose them in the fall

Terrestrial Biomes
Tropical Forests

- Found in South America, Africa, and the Indo-Malayan region near the equator
- The weather is always warm, and rainfall is plentiful (minimum of 190 cm)
- May be the richest terrestrial biome both in numbers of different species and their abundance

Terrestrial Biomes
Tropical Forests

Divided into three levels:

1. Forest floor
2. Understory
3. Canopy

Terrestrial Biomes
Grasslands

- Occur where rainfall is greater than 25 cm but is generally too low to support trees
- Once covered more than 40% of the earth's surface (many areas are now agricultural)
- Grasses are well adapted to a changing environment
- Can tolerate over-grazing, flood, drought, and even fire

Terrestrial Biomes
Prairies

Can be divided into two types:

1. Tall Grass Prairies
    - Requires more rainfall
    - Large tall grasses (>2 m)
2. Short Grass Prairies
    - Require less rainfall
    - Small, short grasses between 5 and 10 cm

Terrestrial Biomes
Savannas

- Grasslands that contain some trees
- Occur in regions where a relatively cool dry season is followed by a hot, rainy season
- Support a large variety of grazers

Terrestrial Biomes

Desert

- Usually found at about 30° latitudes in the Northern and Southern Hemispheres
- Winds in these regions lack moisture
- Annual rainfall is less than 25 cm
- Days are hot and nights are cold

Aquatic Biomes

Freshwater

Inland bodies of water of two types:

1. **Lotic** (flowing water)
    - Rivers
    - Streams
2. **Lentic** (still water)
    - Lakes
    - Ponds

Aquatic Biomes
Lotic

- Characterized by flowing water
- Current is unidirectional
- Species diversity is high, while density is low

Lotic System

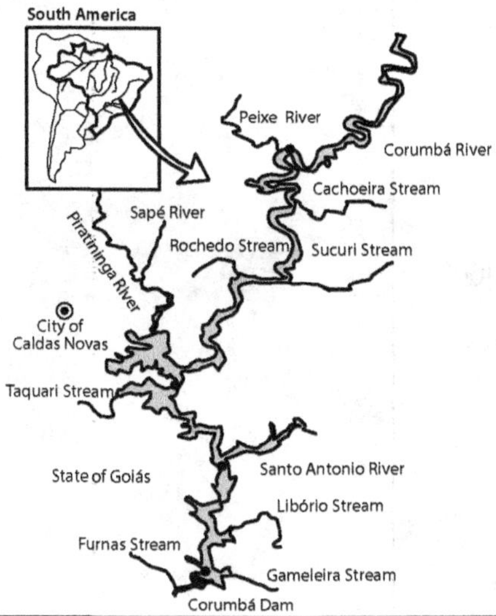

Aquatic Biomes
Lentic

- Ponds
  - Young
  - Shallow
  - No thermal stratification
  - Rooted plants found throughout
- Lakes
  - Old
  - Deep
  - Thermal stratification
  - Rooted plants are absent in the middle

Aquatic Biomes
Thermal Stratification

Summer:

- **Epilimnion**—upper layer, warmed by solar radiation
- **Metalimnion** (thermocline)—narrow, middle layer undergoes rapid temperature drop
- **Hypolimnion**—bottom layer, water is much colder

Oligotrophic	Mesotrophic	Eutrophic
Pond	Marsh	Swamp
Terrestrial	Dystrophic	Bog

Aquatic Biomes
Lake Zonation

- **Littoral Zone**—shallow, warm water near the edge, contains rooted plants and clinging organisms
- **Limnetic Zone**—water that allows for solar penetration throughout, makes up the body of the lake, contains floating plants, as well as floating (plankton) and swimming (seston) organisms
- **Profundal Zone**—area where no light penetrates to this depth, home of crayfish and mollusks
- **Benthic Zone**—bottom substrate (mud)

Aquatic Biomes
Organisms

- **Plankton** (floating)
  - Zooplankton (animals)
  - Phytoplankton (plants)
- **Seston** (swimming)
- **Benthos** (bottom dweller)

Aquatic Biomes
Marine

- **Estuary**—a partially enclosed body of water where freshwater and saltwater meet and mix
- Organisms must be able to withstand constant changes in salinity
- Well over half of all marine fishes undergo part of their development to adulthood in estuaries, which is why they are called ***"The Nurseries of the Sea"***

Aquatic Biomes
Marine

- **Oceans**—play a major role in distribution of the heat in the biosphere
- Air takes on the temperature of the water below, thereby making the winds blow
- Major ocean currents move heat from the equator to cooler parts of the biosphere

Aquatic Biomes
Marine

- Divisions of the ocean:
  - Pelagic Division
    - Neritic Province (above the continental shelf)
    - Oceanic Province (the oceanic basins)
  - Benthic Division
    - Continental Shelf (in sub littoral shelf)
    - Continental Slope (in bathyal zone)
    - Abyssal Plain (in abyssal zone)

Aquatic Biomes
Pelagic Division

- Epipelagic zone (0–120 meters)
- Mesopelagic zone (120–1,200 meters)
- Bathypelagic zone (1,200–3,000 meters)
- Abyssal zone (3,000 meters–bottom)

Aquatic Biomes
Coral Reefs

- Areas of biological abundance
  - Shallow
  - Warm
  - Tropical
- Chief constituents are hard corals (animals with a calcium carbonate (limestone) exoskeleton)
- Areas of intermediate disturbance